STUDIES OF

# THE RUSSIAN INSTITUTE

OF COLUMBIA UNIVERSITY

SOVIET NATIONAL INCOME
AND PRODUCT IN
1937

ABRAM BERGSON

# SOVIET

# NATIONAL INCOME

# AND PRODUCT

# IN 1937

GREENWOOD PRESS, PUBLISHERS
WESTPORT, CONNECTICUT

# THE RUSSIAN INSTITUTE

## OF COLUMBIA UNIVERSITY

The Russian Institute was established by Columbia University in 1946 to serve two major objectives: the training of a limited number of well-qualified Americans for scholarly and professional careers in the field of Russian studies and the development of research in the social sciences and the humanities as they relate to Russia and the Soviet Union. The research program of the Russian Institute is conducted through the efforts of its faculty members, of scholars invited to participate as Senior Fellows in its program, and of candidates for the Certificate of the Institute and for the degree of Doctor of Philosophy. Some of the results of the research program are presented in the Studies of the Russian Institute of Columbia University. The faculty of the Institute, without necessarily agreeing with the conclusions reached in the Studies, believe that their publication advances the difficult task of promoting systematic research on Russia and the Soviet Union and public understanding of the problems involved.

The faculty of the Russian Institute are grateful to the Rockefeller Foundation for the financial assistance which it has given to the program of research and publication.

# PREFACE

This volume is a revised and elaborated version of a two-part study that first appeared in the *Quarterly Journal of Economics* for May and August, 1950. The revision and elaboration embrace a variety of changes, including chiefly the following: (i) A Soviet figure on the aggregate profits for the economy as a whole was previously interpreted as a true net profits figure. For reasons to appear, I now take this figure to represent aggregate profits *before* the deduction of losses covered by subsidies from the government budget. Accordingly, an attempt is made here to allow for subsidies in order to arrive at a true net profits figure. The amount of net profits is used in the calculation of national income as a sum of factor payments and also in the calculation of gross investment as a residual, i.e., as the difference between the sum of factor payments and the total value of the final product allocated to uses other than investment. This revision accounts almost entirely for a reduction in estimated gross investments from 64.6 to 56.1 billion rubles. (ii) As before, in addition to calculating Soviet national income in current rubles, I attempt to appraise the effects on structural aspects of special features of the ruble price system. But the appraisal has now been broadened to include a consideration of subsidies which previously had been passed by. (iii) I now restate and elaborate the brief argument presented previously on the theory of national income valuation. This argument provides the point of departure for the appraisal just referred to of the effects on national income structure of special features of the ruble price system. (iv) The present volume includes a number of appendices that previously were circulated privately in mimeographed form.

I am happy to acknowledge my great indebtedness to Professor A. Gerschenkron for advice and comments concerning both the earlier and the present version. Mr. C. J. Hitch made helpful suggestions at an early stage concerning the theoretic aspects of national income valuation, while in the present version the discussion of this subject owes much also to comments on the *Quarterly Journal of Economics* articles by Professors P. A. Samuelson and J. Marschak, and Mr. N. Kaplan. Professor Carl Shoup kindly subjected the study to a careful and searching review, with special reference to methodological aspects. Mr. Hans Heymann, Jr., provided invaluable assistance in the work on the earlier version generally and substantial aid in the revisions that have now been made to allow for subsidies.

The study was aided at different stages by grants from the Social Science Research Council and from the research funds of the Russian Institute. I greatly appreciate this assistance.

The publication of a revised version of the study that appeared in the *Quarterly Journal of Economics* is with the kind permission of the editors of that journal.

NEW YORK CITY                                                           A. B.

MAY, 1952

# CONTENTS

# TABLES

# SOVIET NATIONAL INCOME
## AND PRODUCT IN
## 1937

# 1 INTRODUCTION

This essay represents in part an attempt to apply to the Soviet Union a novel methodology of national income calculation that recently has been applied to numerous Western countries. This is the methodology of "national economic accounting," [1] involving the calculation of national income as the end product of a series of interrelated sector and global accounts of incomes and their disposition.

I explain subsequently a standard of national income valuation to be called the Adjusted Factor Cost Standard. National income data generally are compiled simply in terms of prevailing money values, and this of course is entirely in order where the concern is only to appraise "monetary" phenomena, i.e., money flows, finance, cost structure, etc. On the other hand, one very often wishes to appraise in some sense or other underlying "real" phenomena, and for this the prevailing money values may or may not be in order. This must depend on their relation to some more ultimate standard appropriate to the given purpose. Adjusted Factor Cost is advanced tentatively here as such a basic standard.

Following the conventional procedure, our accounts for the USSR are compiled in prevailing money values, i.e., in the present case, Soviet ruble prices. While compiling them, however, I also try to appraise their relation to Adjusted Factor Cost. Where possible, I correct for distortions. Without any detailed knowledge of the ruble price system, the reader will be aware that some such inquiry is much in order in a study of Soviet national income. Essentially the concern is simply to clarify the recurring question: "But what do the ruble figures mean?"

The calculations are made only for a single year, 1937, and accord-

---

[1] This designation, suggested in Gerhard Colm, "Experiences in the Use of Social Accounting in Public Policy in the United States," International Association for Income and Wealth, *Income and Wealth*, Series 1, p. 76, seems to be more suitable than the more commonly used "social accounting."

Full identifications of all works referred to in the present study will be found in the Bibliography, pp. 149-156.

ingly can bear only on questions of structure as distinct from trends over time. Also, as will appear, the adjusted Factor Cost Standard is something of a practical expedient; from a theoretic standpoint, I fear, valuation in these terms may raise as many questions as it answers.

On the other hand, this approach to the ruble valuation question may commend itself nevertheless as an interesting alternative to that adopted in previous calculations of Soviet national income by Colin Clark [2] and Julius Wyler.[3] Both these writers abandon altogether ruble prices, and value Soviet national income instead in terms of United States dollars. More recently, Paul Baran [4] has calculated Soviet national income in terms of ruble prices, but without exploring in any detail the limitations.

Furthermore, without regard to the price system, national economic accounts provide an illuminating record of the country's money flows and finance; and in this connection the accounts compiled here may be of value as seemingly the first for the USSR on any scale. All the writers just referred to are concerned more with the total amount than with the structure of national income; accordingly they confine themselves to one or two summary tabulations. Thus, Clark and Wyler focus on a summary tabulation of the national income by use; Baran on a brief account by type of income and use.

Reference has been made to several non-Soviet calculations of Soviet national income. There are, of course, also Soviet data, and brief comment is in order on these at least to indicate the rationale of independent calculations.[5] From this standpoint there are three main

[2] Colin Clark, *A Critique of Russian Statistics;* "Russian Income and Production Statistics," *Review of Economic Statistics,* November, 1947.

[3] Julius Wyler, "The National Income of Soviet Russia," *Social Research,* December, 1946.

[4] Paul A. Baran, "National Income and Product of the USSR in 1940," *Review of Economic Statistics,* November, 1947.

[5] On the Soviet official statistics there is now a considerable body of literature. See in addition to the studies of Clark, Wyler, and Baran already cited, Paul Studenski, "Methods of Estimating National Income in Soviet Russia," in National Bureau of Economic Research, Conference on Research in Income and Wealth, *Studies in Income and Wealth,* Vol. VIII; Paul Studenski and Julius Wyler, "National Income Estimates of Soviet Russia," and Abram Bergson, "Comments," in "Papers and Pro-

features to note. First, ever since the early thirties the Russians have been publishing absolute figures only in terms of the ruble prices of the year 1926–27. Only very summary percentage breakdowns are published in terms of current rubles, and these sporadically.[6] Accordingly, there is hardly any basis here such as national income calculations usually provide to appraise financial aspects. Second, because of certain deficiencies, chiefly centering on the procedure used to value new commodities introduced after 1926–27, the accuracy of the official series in terms of 1926–27 rubles is seriously in question. For what appears to the writer good and sufficient reason, it has been argued that the official series is subject to a substantial upward bias.[7]

Third and last, there is the special nature of the Soviet income concept. Under the influence of what they consider a Marxian distinction between productive and unproductive activity, the Soviet economists confine their measure of national income more or less to the value of material goods produced. Services generally are excluded from national income. Not only personal services, such as domestics, doctors, and teachers, but even government services are omitted. Thus, in terms of Western concepts, the Soviet data are at best only partial measures, and accordingly cannot readily be used for many purposes for which Western data customarily are employed.[8]

---

ceedings of the Fifty-ninth Annual Meeting of the American Economic Association," *American Economic Review*, May, 1947; Naum Jasny, "Intricacies of Russian National Income Indexes," *Journal of Political Economy*, August, 1947; A. Gerschenkron, "The Soviet Indices of Industrial Production," *Review of Economic Statistics*, November, 1947; Maurice Dobb, *Soviet Economy and the War*, pp. 37 ff.; Maurice Dobb, *Soviet Economic Development since 1917*, pp. 261 ff.; Maurice Dobb, "Further Appraisals of Russian Economic Statistics," *Review of Economic Statistics*, February, 1948; A. Gerschenkron, *A Dollar Index of Soviet Machinery Output 1927–28 to 1937*.

[6] In Appendix F, I attempt a reconciliation between the calculations made in this study and the few data the Russians have published in terms of current rubles for 1937.

[7] This question is discussed in most of the writings in note 5. Among the writers cited, Dobb appears to constitute a minority of one in holding that the official data may not be appreciably in error.

[8] In three recently published volumes (*The Soviet Economy during the Plan Era; The Soviet Price System; Soviet Prices of Producers' Goods*), Dr. N. Jasny sets himself the interesting task of calculating Soviet national income in terms of the same standard as is used in the official statistics, i.e., 1926–27 ruble prices, but with a valid valuation of new commodities. I believe there is a good deal of foundation for the

In the last analysis, of course, even an independent calculation must rely on Soviet statistics at some point. This one rests almost entirely on Soviet budgetary and other aggregative financial data in current rubles. Are *these* data reliable? It is, I believe, widely agreed that in respect of reliability Soviet financial statistics in current rubles are on altogether a different plane from Soviet national income statistics in 1926–27 rubles. On the other hand, Professor Gerschenkron's cautions regarding Soviet production data in physical units apply as well to the financial statistics: ". . . the fact remains that very little is known about the exact way in which raw data are collected, reported, collated and prepared for publication. There must be a good deal of distortion that is due to the low educational level in what, after all, is still a rather backward country. There is pressure on individual enterprises in industry to fulfill and overfulfill the plans which may express itself in upward adjustments of reports on output. [In the case of some of the financial data, e.g., wage statistics, presumably the tendency would be rather in the direction of understatement. A.B.]" [9]

But Professor Gerschenkron has expressed very well, too, the rationale of research founded on such deficient information:

assumption implied throughout that the ruble price system was more meaningful on the eve of the five year plans than it was later, so in effect there is still another approach to the problem of valuing Soviet national income. I shall point out in Chapter 3, however, some limitations in the dollar standard of Clark and Wyler that arise because of the differences between Soviet and American preferences and technology. Considering the vast economic transformation in the USSR under the five year plans, the reader will readily see that the procedure used by Dr. Jasny must encounter entirely comparable difficulties.

At one or two later points, I comment briefly on a few of the numerous criticisms which in the studies just cited Dr. Jasny directs against my present study in its earlier version. Also, use will be made of a calculation of Dr. Jasny's on the price level of investment goods. I regret very much that in view of the late date at which Dr. Jasny's studies became available to me it has not seemed advisable to go beyond this and what was said in the previous paragraph to take account here of these volumes.

[9] A. Gerschenkron, *A Dollar Index of Soviet Machinery Output, 1927–28 to 1937*, p. 24. Professor Gerschenkron himself draws a distinction, from the standpoint of reliability, between production data in physical units and data in terms of "various ruble valuations." I believe he must be referring in the latter connection primarily to the official output statistics in terms of 1926–27 and other "constant" rubles. In any event, with possible exceptions to appear, I believe the financial data in current rubles might be considered on much the same plane as the production data in physical units, so far as reliability is concerned.

". . . counsels of pedantic purity are not necessarily the most help-ful ones. At least, the present writer cannot decide to follow them at the price of foregoing acquisition of such little and uncertain knowledge as may result from preoccupation with the admittedly poor and inaccurate Soviet materials." [10]

[10] *Ibid.*, p. 24. Necessarily, insofar as Soviet data of any sort are used by Western scholars, the assumption is that Soviet statistics are not generally falsified in the sense of being freely invented under a double bookkeeping system. It may be in order to summarize briefly the grounds for this assumption as I see them:

(i) To date the data appear to have withstood tolerably well a great many checks as to their internal consistency, e.g., the comparison of household incomes and out-lays in Table 1, p. 18. While not impossible, such consistency reflected in data of diverse sorts and appearing in many different sources might be administratively diffi-cult of attainment under double bookkeeping.

(ii) I believe a careful inquiry would reveal a broad consistency also between the statistical data and other Soviet information on the Soviet economy, as when the release of data on the underfulfillment of the plan in a particular industry has been followed by a report of a change in administrative personnel in the industry con-cerned. This sort of consistency too might be difficult to maintain under double bookkeeping.

(iii) There appears to be a broad consistency too with information of all sorts gathered by foreign observers in the USSR, e.g., Soviet reports of improvements in consumers' goods production are in accord with foreign observations on the supplies available in consumers' goods markets.

(iv) While properly falling under previous headings, the Soviet war experience appears to deserve special mention. In terms of their own statistics, the Russians probably fought the war with an annual steel output averaging during the years 1942–45 little more than 10 million tons. Even taking account of Lend-Lease, one may at least rule out the likelihood of overstatement.

(v) The Soviet policy of withholding obviously is often calculated to mislead, as when the government omitted altogether from the report on the fulfillment of the First Five Year Plan (State Planning Commission of the USSR, *Summary of the Fulfillment of the First Five Year Plan*), data on livestock herds, which had declined drastically under the ruthless collectivization drive. On the other hand, the policy of withholding as so applied appears to be more of an alternative than a complement to a policy of falsification.

(vi) The Russians occasionally release adverse information, e.g., the extent of the decline in livestock herds under collectivization is known from data the Russians themselves published subsequently.

(vii) The Russians published before the German attack a brief summary of their economic plan for the year 1941 (N. Voznesensky, *The Growing Prosperity of the Soviet Union*). There is now available in this country a classified version of this same plan: *Gosudarstvennyi plan razvitiia narodnogo khoziaistva SSSR na 1941 god* (State Plan of the National Economy of the USSR for 1941, published in the United States by the American Council of Learned Societies). The goals in the pub-lished plan check closely with those in the classified version. See L. Turgeon, "On the Reliability of Soviet Statistics," *Review of Economics and Statistics,* February, 1952.

I have said that the published Soviet data appear to be consistent both internally

In research on the Soviet economy, as has just been noted, it is
necessary to reckon not only with deficiencies in the data that the
Russians publish but also with their policy of withholding data from
publication. The government withholds not only strictly military data

---

and with other available information. It is necessary to observe now that there are
important exceptions. But in those cases that have been subject to careful scrutiny
to date, it generally seems possible to understand the inconsistency in terms of diverse methodological deficiencies and without assuming falsification. Probably the
most outstanding case is that of the official national income statistics in 1926–27
rubles, which reveal a rate of growth all out of proportion to the rates shown by
Soviet data on the production of different commodities in physical units. But students
of these data consider that this inconsistency can be explained in terms of such deficiencies as the one mentioned above, p. 5, concerning the valuation of "new"
commodities. For details on this aspect the reader may be referred again to the
various studies cited above, p. 4, note 5. A number of other cases of distortions in
the Soviet data that have been listed by Dr. N. Jasny ("Soviet Statistics," *Review of
Economics and Statistics,* February, 1950) I believe can be explained in similar terms
along lines which Dr. Jasny's own discussion suggests. The writer has analyzed
elsewhere on the same basis still another case concerning wage statistics where the
reliability of the Soviet data is in question ("A Problem in Soviet Statistics," *Review
of Economics and Statistics,* November, 1947).

At the same time, it turns out that the methodological deficiencies generally though
not always operate in the Russians' favor, so the line between these deficiencies and
falsification in the sense of free invention is obviously at best a fine one. Accordingly,
if one wishes to call the methodological deficiencies falsification he is entirely at
liberty to do so. But no one should allow his terminological preferences to blur the
paramount importance of the distinction between such methodological deficiencies
and falsification in the sense of free invention under double bookkeeping. In the
case of free invention, research on the Soviet economy clearly is practically ruled out
at once. In the case of methodological deficiencies, there is at least a core of fact
from which to start and one may hope to detect and even correct the deficiencies.

Why do not the Russians falsify their statistics? I take it as self-evident that the
question is in order. In trying to answer it, my own inclination is to think mainly in
terms of: (i) the probable difficulties in operating a double bookkeeping system on a
national scale without detection; (ii) the possibilities of achieving major propaganda
aims in any case through withholding and perhaps also methodological manipula-
tion.

But having said so much it is necessary to say more. The view on Soviet statistics
that has been set forth necessarily has a more or less provisional character. In the
case of Russia the reliability of official statistics has to be tested and retested and then
tested again. Moreover, if some data appear trustworthy there is no guarantee that
others are likewise.

This last observation, it may be advisable to note, is especially in order today.
While the Russians already were withholding much data on their economy before
the war, they have been notably more secretive since, and during the past year or
so their information releases have been even more arid than previously. Possibly, as
has been suggested above, withholding generally is something of a testimonial to
the reliability of what actually is published. But, with the progressive drying up of

but also data pertaining to the Russian economy generally. The year 1937, however, happens to be a relatively favorable one in terms of the volume of statistical data available, and this is one reason for selecting it for this initial study. Another is that, as will appear, the ruble price system probably was somewhat more meaningful at this time than in some other periods that might have been considered.

The limitations of this inquiry, then, clearly add up to an imposing list; but it may be hoped that the reader who perseveres will be rewarded, nevertheless, with some further illumination on the dimensions of the Soviet economy in a fairly recent peacetime year. Also, students of Soviet economics may find the study of value as an exploratory inquiry in an important field.

It will be useful in various connections to have in mind a few facts concerning the state of the Soviet economy and economic organization in the year studied. In general, 1937 was a prosperous year by

---

information the opportunities for an independent appraisal have been greatly diminished. For this and diverse other reasons, the writer feels that a heightened caution is in order in the use of current Soviet data.

The foregoing remarks summarize in some particulars and amplify in others a view on Soviet statistics which the writer has expressed in various previous studies (*The Structure of Soviet Wages,* p. x; "A Problem in Soviet Statistics"; *New Republic,* Supplement, May 16, 1949), and also in very summary terms in the previous version of the present study. The view I believe is one which is already widely subscribed to in the Soviet economics field. But, let me now explain, it has seemed in order to devote brief space to it, nevertheless, chiefly because of the highly polemical note introduced into discussion of this subject by recent writings of Dr. Jasny (*The Socialized Agriculture of the USSR,* Ch. I; "Soviet Statistics"; "Results of Soviet Five Year Plans," in W. Gurian, ed., *The Soviet Union; The Soviet Economy during the Plan Era; The Soviet Price System*).

What, then, are Dr. Jasny's views? On first sight one gains the impression of a standpoint diametrically opposed to the one set forth here, for Dr. Jasny cites many examples to show that Russian statistics are falsified, yet he consistently omits to explain to the unwary reader (and, let us hope, the equally unwary editor of his books) that his examples are all cases of methodological deficiencies rather than of falsification in the sense of free invention under double bookkeeping.

In fact, Dr. Jasny doubly misleads the reader. For on the one hand, in practice Dr. Jasny, like those who hold the view on these data set forth in previous paragraphs, makes selective use of Soviet statistics—as indeed he must in order to do any research at all in this field. One must conclude that he, like others, considers there is a core of fact from which to start. On the other hand, the reader of Dr. Jasny's writings must gain the impression that I accept Soviet statistics at face value. This is an untruth. Dr. Jasny creates his effect simply by the device of quoting out of context and of concealing from the reader my own extensive criticisms of Soviet statistics.

Soviet standards. The Second Five Year Plan, which it ended, had met with very real success in every sphere; not only in heavy industries, which of course had first priority, but also in agriculture and consumers' goods industries. The First Five Year Plan had brought a catastrophic decline in agriculture, particularly in livestock, mainly as a consequence of the all-out collectivization drive; under the Second Five Year Plan there was a substantial recovery, the grain harvest of 1937 being an all-time record. Under the circumstances, living standards in 1937 probably were higher than in any year since 1928 (the year when the First Five Year Plan was launched), and according to some indications may even have surpassed those of the earlier year.[11] The expansion of heavy industries continued after 1937, but for the consumer prosperity, such as it was, was brought to a close

[11] On this important point, on which there are divergent opinions among students of Soviet economics, I am guided chiefly by scattered Soviet statistics on the production of different kinds of consumers' goods. The Soviet government does not publish any summary measures of the changes in the total volume of consumption and in living standards.

Also, I am referring here to average per capita consumption for the population as a whole. On the eve of the five year plans, I believe, average per capita consumption in the cities must have been well above that in the country. Given this and given also the large shift in the rural population to the cities under the plans and concomitantly the increase in urban bread-winners per family, it is readily seen that with anything like constancy in average per capita consumption of the whole population there very likely would have been an actual decline in urban per capita consumption and a greater decline in real wages. Furthermore, the possibility is open that there might have been some decline in average per capita rural consumption as well.

The previous version of this study, I fear, was open to criticism insofar as the foregoing complexities were not made explicit. On the other hand, the vigorous attack of Dr. N. Jasny on this aspect as overstating Soviet living standards in 1937 (*The Soviet Price System*, pp. 147 ff.) appears subject to the same deficiency. Dr. Jasny cites several items of information to refute the view expressed in the text on living standards, but I believe these items are mainly pertinent to urban and rural consumption and real wages taken separately; the possibility that average per capita consumption for the population as a whole in 1937 approximated that in 1928 is not precluded.

Curiously, while neglecting to reckon with the distinction between these features in criticizing me, Dr. Jasny is himself fully aware of it. Elsewhere (*Soviet Economy during the Plan Era*, pp. 66 ff.) he makes the same distinction that has been made here between consumption for the population generally and for different sectors and refers explicitly to the developments in the USSR which must lead to diverse trends in these different categories.

Dr. Jasny's procedure seems all the more curious in view of the fact that in criticizing me in the *Soviet Price System* he omits altogether to refer to the following

when after Munich the Russians shifted their defense preparations into high gear.

While it prevailed, however, the Soviet government managed, for the first time under the five year plans, to maintain an open market for the sale of consumers' goods. The rationing of consumers' goods had been initiated in 1928 and 1929, practically with the inauguration of the First Five Year Plan. But it was abandoned in 1935 and 1936, and from this time until after the Nazi attack consumers' goods were made available to the Soviet citizens in unrestricted quantities at established prices.[12]

In these same years, as during practically the entire interwar period, there was an open market also for labor. The various forms of labor controls that now prevail in the USSR (e.g., under the Labor Reserve School system) date mainly from 1940. From the period of war communism up to this date, the worker in the USSR was in large measure free to choose his occupation and place of employment at the established wages.[13]

---

calculation that he himself makes in the *Soviet Economy during the Plan Era* (p. 66). The data refer to average per capita consumption for the whole population in constant prices:

| | |
|---|---|
| 1928 | 137 |
| 1937 | 141 |
| 1940 | 126 |

Dr. Jasny is no doubt right in drawing attention to the limitations inhering in these data as a measure of living standards as a result of their inclusion of an increasing volume of factory output that merely took the place of previously unrecorded home processing. On the other hand, his data cited above do *not* include outlays on education and health expenditures. According to his own calculations (*ibid.,* pp. 76 ff.), the inclusion of these outlays would change his per capita consumption figures to the following:

| | |
|---|---|
| 1928 | 148 |
| 1937 | 167 |

Apparently, Dr. Jasny's own data in the *Soviet Economy during the Plan Era* are at least broadly corroborative of my statement on living standards in the text, according to the most obvious construction.

Of course, the reader will understand that I myself do not feel especially committed to an opinion that amounts here only to a prefatory remark; it is hoped there may be an early opportunity elsewhere to examine the important question considered with the care it deserves.

[12] See, however, the comments on the question of the adequacy of supplies in Chapter 4, pp. 63 ff.

[13] The question of freedom of choice of employment in the USSR is discussed in

As of 1937, the collectivization of agriculture in the USSR had been practically completed. In all, 93 percent of the peasant households tilling 99 percent of the land in the hands of such households were embraced within the collective farm system. The collectivized and private households, taken together, tilled about nine tenths of all cultivated land. The balance for the most part was in the hands of the state farms.

The Soviet collective farm often is described as a cooperative organization. Obviously, it has little in common with the farm cooperatives known in the West, but the characterization is not completely true even in a formal legal sense. Reflecting the conflicting pressures under which it was created, the collective farm also embodies legal features of both state and private enterprise. Land, in principle, belongs to the government but is allotted nominally rent free to the collective and also—in plots of strictly limited size—to the individual members for use as homesteads. Large machinery is the property of the government, being made available to the collective farm on a fee basis (the fees are in kind) by the state-owned machine-tractor stations (MTS). Livestock is owned by the collective farm members, in part cooperatively and in part privately. The amount of livestock that can be owned privately by any one homestead is again strictly limited, but in the aggregate this form of ownership constituted in 1937 nearly half of the productive livestock of the USSR. This is now the most important case of private ownership of productive means in the USSR.

Arrangements with regard to the distribution of income are correspondingly intricate; it suffices here to call attention to the most outstanding features. Insofar as the produce itself is concerned, the government receives not only the payments in kind for machine-tractor station services, but additional deliveries, partly on an obligatory basis at very low prices and partly on a voluntary basis at gen-

---

Abram Bergson, *The Structure of Soviet Wages*, pp. 143 ff., and the texts of the Labor Reserve Act and other labor-control legislation enacted in 1940 may be found in Appendix F of this same source. According to the Law of October 2, 1940, establishing the labor reserve system, youths of 14 to 17 years of age are subject to a draft for vocational training, and on completion of this training must work for four years at jobs designated by the government.

erally higher prices.[14] After meeting its obligations to the government, and making provision for seed and fodder, the collective is free also to sell its produce on the retail collective farm market. This is the one completely free market in the USSR, there being no regulation either of supplies or prices. Most of the balance of the produce is distributed in kind to the collective farm members on a "labor-day" basis, i.e., in proportion to the quantity and kind of service rendered. After the money expenses of the collective are met, the money income from the sale of produce likewise is distributed on this basis. The members have at their disposal also the produce of their own homesteads. This, after meeting their obligations to the government, they are free to consume or sell, and the sales may be either to the official procurement agencies or on the retail collective farm market.[15]

I pass by here altogether the question of controls, except to note that at every stage the collective is expected to conform to the government plan, and that this plan is remarkably detailed, containing provisions not only in regard to the quantity of produce but also in regard to timing of operations, etc.

Outside of agriculture, the Soviet economy in 1937 was predominantly in the hands of state enterprise, as indeed it had been since the early thirties. Except for retail trade and particularly small-scale industry, where various types of cooperatives played an important role, state ownership and operation were practically universal.

With regard to the state enterprise, mention should be made here of one interesting feature: financially speaking, it is on its own. Under the so-called system of "economic calculation" (*khozraschët*), an outstanding aspect of the planning system, the state enterprise has its own financial statements and generally is expected to cover its expenses out of its own revenue from sales, or temporarily by loans

---

[14] For purposes of this discussion, the various cooperative procurement agencies, which have an important role in procurements of commodities other than grain, are referred to as government agencies.

[15] The prices paid by the government for voluntary deliveries, while above those paid for obligatory deliveries, generally are well below those prevailing in the collective farm market. Needless to say, this raises a question as to whether the voluntary deliveries are really voluntary. I shall comment on this subsequently.

from the state bank. Accordingly, relations to the government budget are on a net rather than a gross basis. Instead of incorporating all revenues and expenditures of state enterprise, the government budget includes only taxes and allocations from profit, on the one hand, and grants for fixed capital needs (these are the subject of outright grants, rather than loans) and for other special requirements, on the other.

These financial arrangements evidently are similar to those characterizing the public corporation in mixed economies. In respect of the USSR, however, they provide a basis for drawing a distinction which otherwise might be difficult to make: between "government" institutions, properly so-called, and the operating enterprises administering state property. The state budget itself may be accepted as the criterion, and the line drawn according to whether the given organization is or is not on this budget. On this basis, in the economic sphere organizations at the commissariat (since 1946, ministry) level and higher are "government" institutions and organizations below this level, i.e., trusts, combines, and individual factories, are operating enterprises. This, however, is only broadly true. A major exception are the supply and procurement departments of the economic commissariats, which generally operate on the same principles of financial independence as do organizations at lower levels.

But the _khozraschët_ system is rather more complex than the foregoing remarks might suggest and account will have to be taken subsequently of special aspects. On the one hand, the state enterprise is called on to contribute financially to various more or less public activities. These include not only social security, as is true for the private enterprise under capitalism, but also other activities of a more novel sort, including most notably the so-called Soviet trade unions. In the USSR, contributions of the employing enterprise are a major source of funds in the trade union budget.[16] This, of course, is only one of many features that make the Soviet trade unions strange to Western eyes.

On the other hand, many _khozraschët_ organizations have been

[16] I refer here to the time studied. It is believed that the arrangements in question have since been discontinued.

more or less dependent on government subsidies to meet their current expenses. For a time, in the early years of industrialization, subsidies actually were a major factor in the Soviet economy, particularly in the case of basic industrial goods, i.e., coal, steel, machinery, etc. Subsequently, as a result of a price reform initiated in April, 1936, many such products were made self-supporting and subsidies to others were greatly reduced. But, at the time studied, they still were granted on an appreciable scale. Furthermore, after as well as before the reform, subsidies were paid to sectors other than industry, particularly the MTS, in order to compensate for the undervaluation of income in kind at the low government procurement prices for farm produce.[17]

The institutional arrangements now prevailing in the USSR, it should be observed, are much the same as those just described in reference to the year 1937. Hence, insofar as the data assembled here bear on matters of economic organization, they can be taken as broadly pertinent to the system prevailing at the present time.

The accounts that have been compiled in terms of current rubles are set forth in the chapter immediately following. In Chapter 3, I explain the Adjusted Factor Cost Standard, and discuss briefly its rationale and limitations as a norm for the measurement of "real" phenomena. A brief survey of the relation of ruble prices to this standard follows in Chapter 4, along with an appraisal and attempt

[17] Interestingly, during the war the government reverted to the pre-April, 1936, policy and subsidies again were granted on an increasing scale, but in 1949 the rule of profitability was again revived.

Reference should be made here, too, to the special status of state housing enterprises. Until October, 1937, as an exception to the general rule for operating enterprises of an economic nature, these were attached directly to the government budget instead of being on a *khozraschët* basis. On the other hand, the MTS, which previously had been on a *khozraschët* basis, though not fully self-supporting, were attached fully to the government budget in 1938. This latter change meant among other things that the income in kind of the MTS became a revenue item in the government budget.

The question of the status of the Soviet *khozraschët* organization and of its financial relations with the government and other Soviet organizations will have to be considered subsequently at various points in this study, but reference may be made here to some of the more illuminating sources: M. V. Nikolaev, *Bukhgalterskii uchët* (Accounting); S. Glezin, compiler, *Biudzhetnaia sistema Soiuza SSR* (Budget System of the USSR); G. Kozlov, *Khoziaistvennyi raschët v sotsialisticheskom obshchestve* (Economic Accounting in Socialist Society); N. N. Rovinskii, *Gosudarstvennyi biudzhet SSSR* (Government Budget of the USSR).

to correct for the effects of divergencies on our accounts. In the final chapter, I venture to refer briefly to some of the more interesting economic implications of the data compiled. Where it is of interest, comparisons are made in this connection with corresponding data available for the United States.

# 2 NATIONAL INCOME IN CURRENT RUBLES

The national economic accounts that have been compiled in terms of current rubles are shown in Tables 1 to 4. For convenience, I comment first on certain methodological matters, particularly the scope and form of the accounts, the national income concepts used, and one valuation question which it is advisable to consider at this stage, namely, the problem of valuing farm income in kind.

*Scope and form of the accounts.* Here and elsewhere in methodological matters the present study is modeled most nearly after the U.S. Department of Commerce calculations for the United States as recently revised.[1] There are a few divergencies, however, and the comments below are intended to draw attention to these features as well as to one or two aspects which in the present context may not be sufficiently clear from the captions.

*Table 1: Incomes and Outlays of Households.* Households are understood here to include collective farm homesteads but not the collective farm as such. Thus, income received by the collective farm member from his own homestead and also from the collective farm, in money and in kind, is recorded in the account—but income retained by the collective farm for investment and other purposes is not. (The collective farm as such is taken into account along with other organizations in a separate tabulation to be discussed in a moment.)

The introduction into the household account of a category "Transfer outlays" to parallel the category "Transfer receipts" seems to represent an innovation in national income methodology. In the U.S. Department of Commerce and similar calculations only the second heading is used. The rationale of the parallel accounts, however, is readily understood; on the one hand we have income receipts in return for which no goods or services are currently provided and

---

[1] U.S. Department of Commerce, "National Income," supplement to *Survey of Current Business,* July, 1947.

# TABLE I

## INCOMES AND OUTLAYS OF HOUSEHOLDS, USSR 1937 [a]

### (*Billions of rubles*)

| A. INCOMES | | | B. OUTLAYS | | |
|---|--:|--:|---|--:|--:|
| 1. Net income of households from agriculture | | | 1. Retail sales to households | | |
|   a. Wages of farm labor (on state farms, machine-tractor stations, etc.) | 5.6 | |   a. In government and cooperative shops and restaurants | 111.5 | |
|   b. Money payments to collective farmers on labor-day basis; salaries of collective farm executives, premiums | 7.3 | |   b. In collective farm markets | 16.0 | |
|   c. Net money income from sale of farm products | 14.2 | |   c. Total | | 127.5 |
|   d. Net farm income in kind | 32.5 | | | | |
|   e. Total | | 59.6 | 2. Housing (including imputed net rent of owner-occupied dwellings); services | | 19.9 |
| 2. Wages and salaries, nonfarm | | | | | |
|   a. As recorded in current statistical reports | 77.0 | | 3. Trade union and other dues | | 1.1 |
|   b. Other | 21.1 | | | | |
| | | 98.1 | 4. Consumption of farm income in kind; army subsistence | | 35.0 |
| 3. Incomes of artisans; other money income currently earned | | 13.7 | 5. Total outlays on goods and services (1–4) | | 183.5 |
| 4. Imputed net rent of owner-occupied dwellings | | 4.0 | 6. Net savings | | |
| | | |   a. Net bond purchases | 2.9 | |
| 5. Income of armed forces | | |   b. Increment of savings deposits | 1.0 | |
|   a. Military pay | 1.5 | |   c. Other, including increment in cash holdings | 1.5 | |
|   b. Military subsistence | 2.5 | |   d. Total | | 5.4 |
|   c. Total | | 4.0 | | | |
| 6. Statistical discrepancy | | 4.3 | 7. Direct taxes | | 4.0 |
| | | | 8. Total transfer outlays (6–7) | | 9.4 |
| 7. Total income, currently earned (1–6) | | 183.8 | | | |
| 8. Pensions and allowances | | 6.1 | | | |
| 9. Stipends of students | | 2.2 | | | |
| 10. Interest receipts | | .9 | | | |
| 11. Total transfer receipts (8–10) | | 9.2 | | | |
| 12. Total income | | | 9. Total outlays | | 192.9 |

[a] Sources and methods are set forth in Appendix A. Minor discrepancies between calculated sums of items and indicated totals are due to rounding.

on the other hand uses of income in return for which no goods and services are received. The clarification of this parallelism may contribute in a minor way to the understanding of the relation of the household budget to other national income accounts to be discussed later.

*Table 2: Consolidated Net Income and Outlay Account of Government, Social, and Economic Organizations.* For the purposes of this study, government organizations are understood to comprise in the central government the Supreme Court, the Supreme Soviet, and organs attached to them; the Council of Commissars (since 1946, the Council of Ministers), the member Commissariats, and other organs attached to the Council; and, in the republican and local governments, the various comparable organs. The nongovernment organizations, referred to in Table 2, include the various subordinate agencies administering state enterprise—the trusts, combines, and individual enterprises; the consumers' and producers' cooperatives and the collective farms; the trade unions, and, insofar as their transactions have to be taken into account, the Communist Party and other social organizations. The basis for distinguishing government and nongovernment organizations in the economic sphere already has been indicated, i.e., the test is whether the organization is attached directly to the government budget or operates on a *khozraschët* basis.

The account shown in Table 2 has no exact analogue in the U.S. Department of Commerce calculations, but the reader familiar with these calculations will recognize at once that it represents formally, apart from the transfer items, a truncation of the gross national product account compiled by that organization. The result, however, is an essentially novel account,[2] and one which appears of real interest, though perhaps somewhat more so in Soviet than in Western conditions.

The main features of the account to note are as follows: (i) It is a consolidated account in the usual sense that nothing (except the titles of the different income categories) would be changed if all the or-

___

[2] A very near precedent for this type of account is provided by the two tables on the sources and disposition of government finance in N. Kaldor, "The 1941 White Paper on National Income and Expenditure," *Economic Journal,* June–September, 1942, pp. 206 ff.

## TABLE 2

### CONSOLIDATED NET INCOME AND OUTLAY ACCOUNT OF GOVERNMENT, SOCIAL, AND ECONOMIC ORGANIZATIONS, USSR 1937 [a]

*(Billions of rubles)*

| A. INCOMES | | | B. OUTLAYS | | |
|---|---|---|---|---|---|
| 1. Net income retained by economic organizations | | | 1. Communal services | | |
| a. Retained income in kind of collective farms | 2.0 | | a. Health care | 8.9 | |
| b. Retained money income of collective farms | 2.0 | | b. Education | 17.5 | |
| c. Retained profits of state and cooperative enterprise | 7.0 | | c. Other | 1.0 | |
| d. Total | | 11.0 | d. Total | | 27.4 |
| 2. Charges to economic enterprises for special funds | | | 2. Government administration | | 4.4 |
| a. For social insurance | 6.6 | | 3. NKVD | | 3.0 |
| b. For trade unions and special funds for workers' training and education | 2.2 | | 4. Defense | | 17.5 |
| c. Total | | 8.8 | 5. Gross investment, including inventory accumulations, additions to stockpiles, investments in fixed capital, net foreign balance | | 56.1 |
| 3. Indirect taxes; other payments out of incomes by economic enterprises to the government budget | | | 6. Consolidated total value of goods and services disposed of, exclusive of sales to households | | 108.3 |
| a. Taxes on incomes of collective farms | .5 | | 7. Transfer outlays | | |
| b. Payments from profits of state and cooperative enterprises to government budget | 10.0 | | a. Pensions and allowances | 6.1 | |
| c. Turnover tax | 75.9 | | b. Stipends of students | 2.2 | |
| d. Miscellaneous | 4.0 | | c. Interest payments to households | .9 | |
| e. Total | | 90.5 | d. Total | | 9.2 |
| 4. Less: Allowance for subsidized losses | | − 8.0 | | | |
| 5. Consolidated total charges against current product, net of depreciation | | 102.3 | | | |
| 6. Depreciation | | 5.8 | | | |
| 7. Consolidated total charges against current product | | 108.1 | | | |
| 8. Transfer receipts | | | | | |
| a. Net savings of households | 5.4 | | | | |
| b. Direct taxes | 4.0 | | | | |
| c. Total | | 9.4 | 8. Consolidated total outlay, net of sales to households | | 117.5 |
| 9. Consolidated total income | | 117.5 | | | |

[a] Sources and methods are set forth in Appendix B. Minor discrepancies between calculated sums of items and indicated totals are due to rounding.

ganizations referred to were combined financially into one integrated unit. (ii) The account refers to the position of organizations and not households: hence no entry is made on the income side for charges against the current product accruing to households in the form of wages, etc., or on the outlay side for goods and services purchased by the households; only transfers to and from the households are shown. (iii) The account is *net* in the sense that incomes of the organizations engaged in economic operations are shown only after the deduction of expenses incurred in earning these incomes. (iv) For reasons inherent in the logic of double-entry bookkeeping, our account, like the gross national product account of the Department of Commerce, must balance. This is assured by the inclusion of the various transfer items shown. These are the same as those appearing in the household account except that the households' receipts are outlays and the households' outlays, receipts for the organizations.

### TABLE 3
### GROSS NATIONAL PRODUCT ACCOUNT OF THE USSR 1937 [a]
*(Billions of rubles)*

| A. INCOMES | | B. OUTLAYS | |
|---|---|---|---|
| 1. Total income of households, currently earned (Table 1) | 183.8 | 1. Total outlays of households on goods and services (Table 1) | 183.5 |
| 2. Consolidated charges of government, social and economic organizations against current product, net of depreciation (Table 2) | 102.3 | 2. Consolidated total value of goods and services disposed of by government, social, and economic organizations, exclusive of sales to households (Table 2) | 108.3 |
| 3. Net national product | 286.0 | | |
| 4. Depreciation (Table 2) | 5.8 | | |
| 5. Gross national product | 291.8 | 3. Gross national product | 291.8 |

[a] Minor discrepancies between calculated sums of items and indicated totals are due to rounding.

*Table 3: Gross National Product Account.* Essentially this is a highly summary version of the U.S. Department of Commerce account of the same name. It is obtained here simply by consolidating the accounts for households and organizations in Tables 1 and 2. In the consolidation, transfers cancel out.

*Table 4: Gross National Product by Use.* This is a more detailed statement of the outlay side of the gross national product account. The details are again from Tables 1 and 2.

TABLE 4

GROSS NATIONAL PRODUCT BY USE, USSR 1937 [a]

| ITEM | BILLIONS OF RUBLES | PERCENT |
|------|--------------------|---------|
| 1. Consumption of households | 183.5 | 62.9 |
| 2. Communal services | 27.4 | 9.4 |
| 3. Government administration, including NKVD | 7.4 | 2.5 |
| 4. Defense | 17.5 | 6.0 |
| 5. Gross investment | 56.1 | 19.2 |
| 6. Gross national product | 291.8 | 100.0 |

[a] Minor discrepancies between calculated sums of items and indicated totals are due to rounding.

TABLE 5

NET NATIONAL PRODUCT BY ECONOMIC SECTOR, USSR 1937 [a]

| ITEM | BILLIONS OF RUBLES | PERCENT |
|------|--------------------|---------|
| 1. Agriculture | 60.7 | 21.2 |
| 2. Industry and construction | 121.2 | 42.4 |
| 3. Transportation and communications | 16.3 | 5.7 |
| 4. Trade, including restaurants | 37.5 | 13.1 |
| 5. Finance | 1.9 | .7 |
| 6. Services, including government | 38.0 | 13.3 |
| 7. Other | 1.1 | .4 |
| 8. Statistical discrepancy | 9.4 | 3.3 |
| 9. Total | 286.0 | 100.0 |

[a] Sources and methods are set forth in Appendix C. Minor discrepancies between calculated sums of items and indicated totals are due to rounding.

*Table 5: Net National Product by Economic Sector.* This tabulation is constructed in accordance with familiar procedures used in the Department of Commerce calculations and elsewhere.

*Income concepts.* Taking account of the difference in economic systems studied, the concepts of *net* and *gross national product* used in this study appear to be essentially the same as those used by the U.S. Department of Commerce. The concept *net national product,* as used by the Department of Commerce, is of course the same as the concept *national income at market price,* as used by numerous statisticians.

In its calculations, the Department of Commerce also makes use of a concept *national income,* so-called, which corresponds to the concept *national income at factor cost* as used by statisticians generally, i.e., it represents the *net national product* or *national income at market price,* less business taxes and transfers and certain other types of charges included in the value product but not considered as a part of the income of business enterprises and households. This particular concept seems to be of very limited interest in reference to Soviet conditions, where the distinction between business taxes, on the one hand, and the profits of state enterprise on the other is apparently of a rather conventional character.[3] The magnitude of Soviet national income in this sense, however, may readily be computed: Taking as the comparable concept in Soviet conditions the net national product less incomes of economic organizations allocated to special funds, the turnover tax and miscellaneous indirect taxes, we get a total of 197.3 billion rubles.

Coming back to the concepts of net and gross national product, it may be advisable to refer explicitly to several aspects:

(i) A recurring question in national income literature concerns the treatment of "government," particularly whether this category is to be classified wholly as a "final product," or whether alternatively it is to be classified in some part as an "intermediate product." Following the Commerce Department, I use the former approach here. At the same time, "government" is understood in this connection to

---

[3] Further details on these two types of income are presented in the section on income and outlay categories, pp. 33 ff., and in Chapter 4 in the section beginning on p. 56. Insofar as in its latest revision the Department of Commerce draws a line between corporate income taxes, on the one hand, and business taxes on the other, the precise significance of its concept of national income even in regard to capitalist conditions is blurred.

comprise a variety of activities of a more or less familiar sort that in
the USSR are supported by appropriations from the government
budget: education, health care, and the like; government administration as such; internal security; defense; and certain other special
measures, such as industrial resettlement, geological surveys, plant
disease control, etc. This is not the place to debate this methodological
question concerning "government," and there is little point in trying to defend the particular application made here of the Commerce
Department's approach. But it is readily granted that the approach
is rather dubious in the case of the special measures just referred to,
particularly since under the Soviet *khozraschët* system no charge appears to be made in the books of state enterprise for depletion, and
furthermore no attempt is made here to correct for this deficiency.[4]

(ii) As has been mentioned, the Soviet *khozraschët* organization
is called on to support financially various semipublic activities. In
several instances such contributions are charged to the organization
as expenses before profits. But except insofar as they lead to mere
transfer of income to households, the activities in question clearly
ought to be considered as a final product like the activities of "government" rather than as an intermediate product. As far as is feasible,
they are so treated here. We refer to contributions to the social security system, to the trade unions, and to certain funds for workers'
training and education.

(iii) It was mentioned, too, that at the time of this study many
*khozraschët* organizations were subsidized from the government
budget. Actually, according to the practice prevailing at the time of
this study, the government appropriated funds to *khozraschët* organizations for a variety of purposes, including chiefly investments
in fixed capital; investments in "own" working capital (representing
minimum working capital requirements as distinct from bank fi-

---

[4] In the latest revision of its income data, the Commerce Department itself aban-
dons the allowance for depletion that appeared in previous calculations. But this is
on the ground that in calculating gross capital formation no allowance is made for
the value of new discoveries. Insofar as the government outlays on geological prospecting are treated here as a final product they seem to find their way ultimately into
gross capital formation, and hence might be construed as representing the value of
new discoveries at cost.

nanced working capital for seasonal and other special needs); the financing of special measures of the sort referred to above, e.g., geological surveys, insofar as these were delegated to *khozraschët* organizations; and to cover losses. It goes without saying that insofar as investments in fixed and working capital actually materialize, they become a final product in this study. In the interests of consistency, I treat likewise the appropriations for special measures but, of course, activities of the *khozraschët* organization reflected in expense charges on its books have to be considered as intermediate products even though losses are incurred and are subsidized by the government. I shall use the term "subsidies" here to refer only to the last kind of appropriation.[5]

(iv) Following the Commerce Department, I include in the income and consumption of households an allowance for the subsistence of the armed forces; and interest on government debt is treated as a transfer payment rather than earned income.

As is already implied, on the whole question of delimiting the scope of intermediate products, the point of departure is the

---

[5] But our classification of the appropriations to economic organizations, while essentially that used in the Soviet budget itself, is something of an oversimplification; in any event reference ought to be made here to the appropriations compensating procurement agencies for extra payments to farmers for deliveries of certain agricultural products (grain and sunflower seed are the only ones known to have been covered at the time studied) above the obligatory norms and for certain procurement operations. Formally these appropriations are similar to those for special measures referred to in the text, and it is believed they are classified on this basis in the Soviet budget; but obviously they come to the same thing as subsidies and are so treated here.

These arrangements, by the way, are rather puzzling, in view of the fact that for many agricultural products the government has found it convenient to deal with the problem of multiple procurement prices simply through differential sales taxes on procurement agencies or alternatively by refunds from such taxes. Possibly the explanation of the budget appropriations is the desire to establish firmer central budget control over tax collections on some products.

I refer subsequently to the question of Soviet budget classification. On the various arrangements to deal with the problem of multiple agricultural procurement prices, see "Instruktsiia NKF SSSR . . ." (Instructions of the People's Commissariat of Finances USSR . . .), December 9, 1937, *Finansovoe i khoziaistvennoe zakonodatel'stvo* (Financial and Economic Law), 1937, No. 36, p. 9; K. N. Kutler, *Gosudarstvennye dokhody SSSR* (Government Incomes of the USSR), pp. 88 ff., 112 ff.; A. K. Suchkov, *Dokhody gosudarstvennogo biudzheta SSSR* (Incomes of the Government Budget of the USSR), pp. 32 ff.

*khozraschët* system. In general, activities recorded as expenses in the books of Soviet enterprise are treated as intermediate in this study. Needless to say, this is a necessary expedient; with the available information there is hardly any basis to question Soviet accounting practice except in the outstanding cases referred to above. Actually, this practice is generally of a rather conventional sort, but there are special features. For example, at the time studied, allocations to reserves for bad debts were charged not as an expense before profits but as a debit to the profits account; and so likewise were such curious items as "Outlays for experiments not yielding results" and "Losses from production risks not envisaged in the plan." [6]

*Valuation of farm income in kind.* Farm income in kind is valued here at the average of realized farm prices. This procedure, of course, is the conventional one in national income calculations; but from what has been said already about the Soviet price system, it must be apparent that it has a rather special meaning under Soviet conditions. The average of realized farm prices here is an average of prices prevailing in quite diverse markets, including the low prices paid by the government for obligatory deliveries, the somewhat higher prices paid by various procurement agencies for additional wholesale deliveries on a voluntary basis, and the still higher prices prevailing in the retail collective farm market. Furthermore, in the USSR there is a spread between wholesale farm delivery prices and the retail prices of foodstuffs as a result not only of the costs of processing, transporting, and distributing costs, but also of the remarkably high sales (in Soviet terminology "turnover") taxes levied on

[6] Aside from the omission of adjustments for special features of the Soviet accounting system, it should be noted that application of the Department of Commerce's methodology was limited generally by a lack of data. Thus, in contrast to the Department of Commerce procedure, no adjustment is made here for inventory valuation. Also, it is not clear whether financial institutions in the USSR render to households any appreciable amount of free service; in any event no allowance is made here for the "imputed interest" in this sense, which appears in the Department of Commerce calculations. Other minor discrepancies of a like sort could be cited.

Concerning the status in the accounts of the *khozraschët* organizations of the different kinds of outlays referred to in preceding paragraphs, sources and further details will be set forth subsequently. On the *khozraschët* system generally, see the references cited above, p. 15, note 17.

foodstuffs along with other consumers' goods. Details on this will be provided subsequently.[7]

This is not the place to comment in any detail on the question of the validity of the "realized-farm-price" valuation procedure in reference to Western economic conditions. While it is of course open to the criticism that it gives a misleadingly low impression of farm in relation to nonfarm real income, there is something to say for it as a basis for measuring the value of the services actually produced by agriculture, and hence the national income of this sector in this sense. And evidently the case is made stronger if as usually is done the convention is adopted that home processing activities, including those on the farm for personal consumption, are to be excluded from national income produced. In relation to the USSR, the indicated procedure obviously is highly dubious, from any point of view, and it is accordingly regarded for the present only as a statistical expedient. We shall be in a better position to appraise it later, however, when the question of the valuation of Soviet national income as a whole is taken under consideration.

### INCOME AND OUTLAY CATEGORIES: GENERAL

Continuing with the explanation of Tables 1–5, I now comment briefly on the income and outlay categories tabulated in the different accounts. These comments are intended in part to indicate, insofar as the captions are not adequate for the purpose, the nature and scope of the different categories, and in part to convey to the reader a general impression of the nature and limitations of the statistical data presented. Details on the sources used as well as on various independent calculations and estimates referred to are presented in Appendices A–C.

Attention is turned first to the household account, then to the consolidated account for organizations, and finally to the tabulation of

[7] The spread is reduced somewhat, however, insofar as "realized prices" are construed here as the average of farm prices, not only net of sales taxes but also gross of subsidies on agriculture, particularly the MTS and state farms. As was indicated earlier, the low procurement prices which the government allowed the MTS and state farms for their produce necessarily meant subsidized losses for these agencies.

national income by sectors. The various income and outlay categories in the remaining two tabulations compiled are taken directly from these three and require no separate explanation.

INCOME AND OUTLAY CATEGORIES: THE HOUSEHOLD
ACCOUNT (TABLE I)

*Wages of farm labor (on state farms, machine tractor stations, etc.).* According to current Soviet statistical reports, the total wages paid employees on state farms and machine tractor stations amounted to 5.3 billion rubles in 1937. Besides this, account is taken here of the wages paid to a limited number of nonmember employees hired by the collective farms, presumably mainly on a seasonal basis. These, according to a Soviet estimate,[8] amounted to .3 billion rubles.

*Money payments to collective farmers on a labor-day basis: salaries of collective farm executives; premiums.* According to a Soviet source, the payments on a labor-day basis amounted to 6.8 billion rubles in 1937.

*Net money income from sale of farm products.* According to calculations based on scattered Soviet data, the gross money income from sales of farm produce realized by the collective farm households, as well as the few remaining independent peasants, amounted to 16.6 billion rubles in 1937. This includes revenue from sales to government and other procurement agencies as well as from the retail collective farm market. A more or less arbitrary sum of 2.4 billion rubles is deducted from the gross sales revenue to allow for the money expense incurred by the collective and other farm households in farm operations.

*Net income in kind.* This is intended to represent the amount of produce which is left at the disposal of the collective and other farm households (including workers' households with garden plots) after

---

[8] The estimate is taken from one of two Soviet studies of household incomes and outlays which are cited in Appendix A but to which specific acknowledgment ought to be made here. The studies, by N. S. Margolin, contain not only a considerable amount of statistical data (unfortunately for our purposes complete tabulations are presented only for the years 1934 and 1938, and these only by broad categories and in percentage terms) but also careful descriptions of the different items tabulated. On both accounts they were most helpful to the present study.

sales are made and production expenses covered; and which accordingly is available either for investment in the expansion of household farm operations or for consumption. As has been explained, income in kind is valued at average realized farm prices. The figure cited in the table is calculated from diverse Soviet data, including mainly statistics on the aggregate value of marketed produce in terms of these prices; on the share of different crops marketed, which averages somewhat more than a third (from the foregoing, a figure on the total harvest in terms of realized farm prices is computed); and on agricultural production expenses, which again come to something more than a third of the total harvest. Because of possible errors in our calculations and no doubt also in the underlying Soviet data, the result is subject to a sizable margin of error.

*Wages and salaries, nonfarm: As recorded in current statistical reports.* The statistical series on the wage bill, from which is taken the item for 1937 cited here (after the deduction of farm wages), was compiled by TSUNKHU, the leading Soviet statistical agency.[9] This TSUNKHU series has been widely interpreted as being a comprehensive figure for the whole Soviet economy, but it appears that this is not the case, and that the figure cited does not include: (i) certain premium payments, for example, payments out of the so-called Fund of the Director; and (ii) the wages paid workers in certain local industries, in secondary lines of activity (factory, restaurants, etc.) in particular plants, and in numerous other spheres.[10] Besides wages and salaries, properly so-called, the TSUNKHU figure includes, though it does not necessarily cover fully, literary and other honoraria.

*Wages and salaries, nonfarm: Other.* This item is calculated on

[9] As appears in the Appendices, most of the figures taken here from current Soviet statistical reports were compiled by TSUNKHU (*Tsentral'noe Upravlenie Narodno-Khoziaistvennogo Uchëta,* Central Administration of National Economic Accounting), which at the time considered in this study was under the State Planning Commission. In a reorganization of the Soviet statistical services of 1939, TSUNKHU, which apparently had enjoyed previously a considerable measure of autonomy, was integrated more fully with its parent agency and renamed TSU, which stands for *Tsentral'noe Statisticheskoe Upravlenie* (Central Statistical Administration). But there was still another reorganization in 1948, with the result that TSU is now no longer under the State Planning Commission but directly under the Council of Ministers.

[10] See Abram Bergson, "A Problem in Soviet Statistics."

the basis of scattered Soviet data, the scope of which unfortunately is not entirely clear. It is believed to include not only the items omitted from the figure—cited above from current statistical reports—which refers to free labor, but probably also the wage paid penal labor.[11] Apparently wages are paid such workers in the USSR, though the rates as well as the numbers of such workers are conjectural.[12]

*Incomes of artisans; other money income currently earned.* The magnitude of this item is calculated independently from various Soviet estimates. It includes the earnings of both cooperative and independent artisans (shoemakers, tailors, etc.) and, it is believed, the earnings of various other independent occupations, such as draying, gold mining, and domestic service. Certain other money income currently earned, such as rents from privately owned housing, and some money income not currently earned, such as the receipts from the sale of secondhand clothing, may also be included.

*Imputed net rent of owner-occupied dwellings.* This rule of thumb estimate is intended to represent the value of the services of dwellings—mainly in rural areas—that are occupied by private owners. The services are arbitrarily valued at the very low rental rates fixed by the government for state-owned apartments in the cities.

*Income of armed forces: Military pay.* A very rough independent estimate of the total money paid to members of the armed forces, which are taken to number 1.75 million men in 1937.

*Income of armed forces: Military subsistence.* It is assumed that at prices paid by the defense commissariat the value of food and clothing provided the armed forces amounted to about 1,500 rubles per man in 1937. In working-class families, the average annual expenditure on food and clothing amounted at retail prices to about 740 rubles per head (including children) in 1935, and probably exceeded 1,000 in 1937.

*Statistical discrepancy.* This represents the difference between the sum of accounted for incomes and the sum of accounted for outlays.

[11] *Ibid.*, pp. 240–41; also Harry Schwartz, "A Critique of 'Appraisals of Russian Economic Statistics,'" *Review of Economic Statistics,* February, 1948, pp. 38–41.
[12] See Appendix C, pp. 122 ff.

The indicated error, it is believed, must affect the figure on "total income currently earned" (Table 1, A, item 7) rather than "transfer receipts" (Table 1, A, item 11).

*Pensions and allowances.* This figure, from a Soviet source, is believed to include not only the pensions and allowances received under social insurance arrangements, but also payments to mothers of many children and military pensions.

*Stipends of students.* As estimated on the basis of information in Soviet sources, this is intended to represent the total amount of living and other allowances paid to students in all educational institutions. As of 1937, tuition was free to all in the USSR.

*Interest receipts.* This allows for a return of 4 percent on the 19.5 billion rubles of government bonds and of 3 percent on the 4.0 billion rubles of savings deposits that were held by private persons in the USSR in 1937.

*Retail sales to households: In government and cooperative shops and restaurants.* According to current Soviet statistical reports, the total turnover of these outlets amounted to 125.9 billion rubles. In order to obtain a figure representing sales to individuals, this is reduced by 11.5 percent which, according to a Soviet sample investigation, represents the proportion of the turnover of these outlets that consists of sales to enterprises and institutions. The cooperative shops are those run by consumers' cooperatives. In 1937 they functioned almost exclusively in rural localities.

*Retail sales to households: In collective farm markets.* According to a Soviet estimate, the total sales of collective farms as such, and of collective and other farm households on the retail collective farm market amounted to 17.8 billion rubles. This figure likewise is reduced to allow for sales to institutions and enterprises which, according to the same Soviet investigation referred to above, amounted to 10 percent of the turnover.

*Housing (including imputed net rent of owner-occupied dwellings); services.* In addition to the imputed net rent of owner-occupied dwellings, this item includes, on the basis of Soviet estimates, the money outlays of households on rents, utilities, transportation, entertainment and, it is believed, personal services such as shoe repairing, tailoring, and presumably domestic services.[13]

*Trade union and other dues.* This item, calculated from Soviet data, is intended to be a comprehensive figure including, in addition to trade union dues, dues paid to the Communist Party and other organizations. Trade union dues alone probably amounted to about half a billion rubles.

*Consumption of farm income in kind; army subsistence.* See the comments on the corresponding items on the income side. Presumably part of the farm income in kind was invested by the household rather than consumed, so that the inclusion of the total figure here tends to overstate the consumption and understate the savings of households.

*Net savings.* The figures for bond purchases and the increment of savings deposits are calculated from a Soviet tabulation of the yearly totals of these items. In the case of bond purchases, the government budget indicates a figure about 1.3 billion rubles larger than that in Table 1. The reasons for this discrepancy are not clear. Possibly there is a difference as between the tabulation of outstanding debt and the government budget in regard to the basis on which the debt is computed: i.e., bond subscriptions being recorded in one case and cash payments for bonds in the other, or vice versa. The increment of cash holdings is fixed at 1.0 billion rubles in the light of Soviet data on cash in circulation compiled by A. Z. Arnold for the period through 1936. Also included in other savings is an arbitrary sum of .5 billion rubles, which is intended to allow for savings in the form of initiation payments to cooperatives, life insurance (such insurance is available in the USSR), and such other forms of money savings as may exist.

The total net savings as computed are understated, insofar as they do not include: (i) investments by farm households out of their income in kind, which is included instead under consumption in kind—this investment, it is believed, may well have amounted to a

---

[13] It is not clear whether it includes also expenditures on housing repairs. The figure on imputed net rent from owner-occupied dwellings, 4.0 billion rubles, is the same as that used in the tabulation of incomes. Logically, I suppose, the appropriate figure to use on the outlay side would be one representing gross rather than net rent.

billion or so rubles in realized farm prices; and (ii) investments in housing by both the nonfarm and farm population [14]—in view of the very limited extent of private housing construction in the USSR at the time considered, this presumably would amount to a very small sum.

*Direct taxes.* This Soviet budget figure includes the revenue from: (i) the "Income Tax," so-called, levied mainly on wages, salaries, honoraria, incomes of artisans; (ii) the "Cultural and Housing Construction Tax," for all practical purposes another income tax but applying to the farm as well as nonfarm population and apparently with the uses of the revenues restricted to special purposes; (iii) the "Agricultural Tax," levied as a fixed amount per farm household, the amount however varying with the region; and (iv) certain other taxes and fees levied on the population.

INCOME AND OUTLAY CATEGORIES: THE CONSOLIDATED NET
INCOME AND OUTLAY ACCOUNT OF GOVERNMENT, ETC.
(TABLE 2)

*Retained income in kind of collective farms.* This is a rather arbitrary estimate of the amount of farm income in kind consisting of investments by collective farms as such—as distinct from member households—in livestock herds, seed and fodder funds and other capital. The farm income in kind shown above in the household account is net of this sum.

*Retained money income of collective farms.* This sum, calculated from Soviet data, represents the amount of current money income that collective farms retain rather than distribute among their member households. The retained income is used to finance capital investments, repay debts, and to finance various communal projects, such as nurseries, etc.

*Retained profits of state and cooperative enterprises.* According to a Soviet source, the total profits earned by government economic enterprises and producers' and consumers' cooperatives amounted to 17 billion rubles in 1937. Of this sum, Soviet budget data indicate

[14] If, as is uncertain, this item is covered elsewhere in the tabulation, presumably it would be under outlays for housing.

that 10.1 billion rubles were transferred to the government budget while 6.9 billion rubles were retained by the operating enterprises as an addition to their own capital. Soviet figures on profits of the sort just cited usually are interpreted as true net profits figures, and the one cited here was so interpreted in an earlier version of this study; but it is now believed that the figure of 17.0 billion rubles represents instead total profits prior to the deduction of losses covered by government subsidies. Accordingly, allowance is made elsewhere in Table 2 for such subsidized losses (item 4).[15]

*Charges to economic organizations for special funds: For social insurance.* This is the aggregate amount, as indicated by a Soviet source, of premiums collected from economic organizations for the financing of the government social insurance program. The entire amount of the premium is an expense of the economic organization; the insured, nominally, does not bear any part of the cost.

*Charges to economic organizations for special funds: For trade unions and special funds for workers' training and education.* The contributions to the trade unions, which on the basis of Soviet data are calculated at 1.2 billion rubles, are not to be confused with membership dues. The contributions are payments which by law the employing establishment itself must make to the trade unions for their support, the payment being a charge against income before profits. The employing establishment is required also to maintain, by charges against income before profits, certain funds for the support of training and educational activities among the workers, including the factory-apprentice school, advanced technical courses, stipends for workers on leave for study at outside educational institutions, and so on. The amounts so allocated out of current income totaled 1.0 billion rubles in 1937.

*Taxes on incomes of collective farms.* This represents, according to Soviet budgetary data, income taxes paid by the collective farms as such, as distinct from the households.

[15] Also, the retained profits as calculated here represent profits before allocations to the so-called Director's Fund, established in April, 1936. Since premiums paid to workers out of this fund supposedly are already included in wages, this means that there is a slight amount of double counting in our calculation of the gross national product.

*Payments from profits of state and cooperative enterprise to government budget.* See above, the comments on retained profits.[16]

*Turnover tax.* This famous tax, on which I cite a Soviet budget figure, is in effect a sales tax. It is levied at widely diverse rates on all sorts of products, but falls almost in its entirety on foodstuffs and other consumers' goods. In general, the principle is followed of levying the tax at only one economic stage for each product, i.e., on the manufacturer, on the wholesale distributor, or on the retailer. Apparently, however, no consistent practice is observed in regard to the taxation of products that are processed from taxed products, e.g., while there is a heavy tax on grain at the procurement stage and none on bread at any stage, there is a tax on flour sold at retail.[17]

*Miscellaneous indirect taxes.* This is a more or less arbitrary estimate of the total of a variety of indirect taxes, fees, and charges, including among others the so-called tax on "noncommodity operations," levied on organizations engaged in custom repair work, autobus transport, etc.; certain stumpage fees levied on organizations using the state forests; fees levied on organizations engaged in commercial fishing; customs duties; urban real estate taxes and ground rents levied on commercial organizations (there are such in the USSR); certain local fees levied on farmers selling in collective farm markets; notarial fees, etc.

*Allowance for subsidized losses.* According to information in a Soviet source, appropriations from the government budget for "outlays on mastering" (*raskhody po osvoeniiu*) were planned to total 6.5 billion rubles in 1937. This designation is an abbreviation for "outlays on mastering of production"; which in turn, at the time studied, was an established Soviet euphemism for subsidies. But it is believed that this total is not entirely comprehensive of subsidies as understood here; more particularly it probably does not include certain appropri-

[16] The term "deductions from profits" (*otchisleniia ot pribylei*), by the way, is used in Soviet financial literature to refer to transfers made by state enterprise only, the more familiar term "income tax" (*podokhodnyi nalog*) being used to refer to the transfers made by cooperative enterprise.

[17] As these brief comments may suggest, the turnover tax is a quite complex fiscal device. Further details on it may be found in Kutler, Ch. III; Suchkov, Ch. II.

ations to procurement agencies previously referred to.[18] Accordingly, I have raised the budget total by an arbitrary sum of .5 billion rubles in order to allow for these appropriations. Finally, the budget total represents a goal. I assume here that the actual payments exceeded the goal by 1.0 billion rubles. Apparently, subsidies in the USSR are intended to compensate only for planned losses. But presumably the plan is subject to revision in the light of changing circumstances. On the other hand, the indications are that any extra appropriations made in 1937 would have been rather limited.[19]

*Depreciation.* The figure cited for this item, taken from the Soviet financial plan for 1937, is believed to represent the estimated sum of the debits for depreciation to be recorded in the books of Soviet enterprise in that year.

*Transfer receipts.* See above (pp. 32–33), the comments on the transfer outlays of households.

*Communal services.* These outlays include, besides the expenditures of the government, comprising the bulk of the total, such expenditures as those of the state factories on workers' training and education, to which reference already has been made; and expenditures for the upkeep of sanatoria, rest homes, and other health measures financed by social insurance funds. The figures on health care and education, which are calculated from scattered Soviet information, are net of outlays on capital construction and transfer payments (stipends, pensions, and allowances). The figure cited on "Other" outlays is an arbitrary sum intended to allow for outlays on such activities as physical culture, social insurance administration, etc.[20]

*Government administration.* This is a Soviet budget figure representing the aggregate expenditures for the upkeep of all government

[18] See above, p. 25, note 5.

[19] Insofar as there are unplanned losses, the alternative to extra subsidies would be bank credit creation. But for present purposes, except where the budget itself is in question, these two seem to come to much the same thing; accordingly, the allowance of 1.0 billion rubles for extra subsidies might be viewed alternatively as an allowance for such bank financed losses.

[20] In the absence of data, it may be assumed that this account allows also for the amount of the excess, if any, of the administrative costs of trade union, party, and other social organizations over and above the aggregate receipts of these organizations from membership dues. According to strict logic, an allowance should be made for this sum somewhere on the outlay side of Table 3.

organs (in the sense indicated earlier), exclusive of the all-union commissariats of defense, navy, and internal affairs.

*NKVD.* The exact scope of this item, for which again a Soviet budget figure is cited, is not known. In addition to its internal security activities, the NKVD maintains military detachments which reportedly were used in the recent war. If so, the outlays for this purpose evidently should be considered as supplementing the defense expenditures referred to below.

*Defense.* The Soviet budget figure on defense cited here is deficient also insofar as it does not include some or all defense plant construction. According to an established Soviet administrative practice, munitions production in the USSR is the responsibility not of the ministries charged with actual military operations but of ministries specialized to this end.[21] The budget figure on defense expenditures represents the appropriation for the ministries charged with military operations only. Insofar as these ministries had to settle with the munitions ministry for the munitions transferred to them, the budget figure for defense expenditures presumably would cover the costs of such munitions. Insofar as defense plant construction was undertaken mainly by the munitions ministry, however, this would not have been included in the budget figure on defense. According to the 1937 budget forecast, the government appropriation to finance the investments of the munitions ministry was to amount to 2.3 billion rubles in that year.

*Gross investment, including inventory accumulations, etc.* This item is calculated here as a residual (i.e., as the difference between the sum of all recorded charges and the sum of all outlays other than investments), and accordingly is subject to the net resultant of errors in estimates of all other items. An independent calculation based on scattered Soviet data on major investment components, i.e., fixed capital, working capital, etc., suggests that I may be underestimating the total by a few billion rubles. Insofar as defense plant

[21] Thus, in 1937, while military operations were the responsibility of the Commissariat of Defense, so-called, and the Commissariat of the Naval Fleet (created in July, 1937), munitions production was largely if not entirely the responsibility of the Commissariat of Defense Industry.

construction is not included under defense above, it necessarily is included in the gross investment calculated as a residual. Government-supported prospecting, resettlement, plant disease control activities and the like presumably find their way, too, into this residual category.

*Transfer outlays.* See above (p. 31) the comments on the transfer *receipts* of households.

## THE NATIONAL PRODUCT BY ECONOMIC SECTOR (TABLE 5)

The data in Table 5 on the industrial origin of the Soviet national product must be considered as very rough indices of the magnitudes involved rather than as in any sense accurate measures. Because of deficiencies in the available information, it was possible to establish the industrial origins of some 10 to 15 percent of the national product only on the basis of rule-of-thumb estimates, which may be subject to an appreciable margin of error.[22] Also, as is indicated in the table, another 3.3 percent of the national product is not accounted for at all by the data on the various economic sectors listed.

Mainly because of the nature of the underlying Soviet data, the sector on Industry in the table has something of a mixed character. It includes not only mining, manufacturing, and electric power but also lumbering and fishing.[23] The latter, of course, are industries of some consequence in the USSR. As of 1935, there were 1,300,000 employees engaged in lumbering and related occupations, and 128,000 in fishing. At the same time, there were 7,466,000 employees in mining, manufacturing, and electric power, and 2,204,000 in construction.[24]

---

[22] This includes the 21.1 billion rubles of wages which, as has been indicated, is believed to represent the wages paid penal workers as well as various categories of free workers not covered in TSUNKHU reports (Table 1 item, "Wages and salaries, nonfarm: Other"). For purposes of allocating this item by industrial sector, I assumed arbitrarily that one sixth of the total or 3.5 billion rubles, represents the wages of penal workers and the remaining five sixths, or 17.6 billion rubles the earnings of free workers. At a rate of pay equal to one half that of free workers, this would allow for a penal labor force of 2.5 millions. This would be the relation in earnings only if the value of subsistence is included; the actual money wages of penal workers are far lower. See Appendix C, pp. 122 ff.

[23] Also included under "Industry" are local public services such as electric stations, gas and water works, etc. Local transport services are intended to be included under "Transport and Communications."

[24] TSUNKHU, *Trud v SSSR* (Labor in the USSR), pp. 10–11.

In Table 5 construction is grouped with industry under one heading.

Besides government, the sector "Services" is intended to include housing, education, health care, cultural facilities, social organizations, and all personal services, such as those of domestics, barbers, etc. The precise nature of the "Other" branches of the economy referred to in the table is not known; presumably the industries included here should come under one or another of the other sectors listed.

### THE GOVERNMENT BUDGET

For all practical purposes a government budget is itself an economic account and from various standpoints it is of interest to have at hand here, along with the accounts that have been compiled by the writer, the Soviet government budget for the year studied. This is shown in Table 6. Reference here is to the so-called "State Budget of the USSR," a consolidated budget of the all-union, republican, and local governments.[25] For the purpose of clarifying the relation of this account to others presented in this study, I have taken the liberty to revise the classification and, on the basis of independent data, to elaborate certain of the categories that appear in the usual Soviet version of this budget. The nature of the revisions can be ascertained at once by comparing the budget shown with that in the Soviet sources cited in Appendix D. The figures cited, of course, represent realized rather than projected magnitudes.

The nature of the "statistical discrepancy" referred to already has been explained. In incorporating the social insurance budget in the government budget, I am following the practice initiated in the USSR in 1938. Prior to that time, only the current surplus of revenues from the social insurance payroll tax over expenditures of the social insurance system was included in the government budget. The contents of the residual revenue category in the budget are rather diverse.

[25] This is according to the usage prevailing since 1939. Prior to 1939, the "State Budget" (*Gosudarstvennyi Biudzhet*) comprised only the all-union and republican accounts, while the "Consolidated Budget" (*Svodnyi Biudzhet*) represented these together with the local accounts. See A. Baykov, *The Development of the Soviet Economic System*, pp. 388–89.

For purposes of this study, it is assumed that it consists of revenues on capital account, such as might be realized from capital transactions with the banking system, sales of assets, etc.

TABLE 6

GOVERNMENT BUDGET OF THE USSR 1937 [a]

(*Billions of rubles*)

| A. REVENUES | | | B. EXPENDITURES | | |
|---|---|---|---|---|---|
| 1. Direct taxes | | 4.0 | 1. Interest charges on debt | | |
| | | |   a. To households | .8 | |
| 2. Net borrowing | | |   b. To savings banks | .2 | |
|   a. From households | 2.9 | |   c. Other organizations | .2 | |
|   b. From savings banks | 1.1 | |   d. Total | | 1.1 |
|   c. From other organizations | −1.8 | | | | |
|   d. Total | | 2.2 | 2. Pensions and allowances including those paid by social insurance system | | 6.1 |
| 3. Statistical discrepancy between budget data on current loan transactions and other data on outstanding debt | | 1.3 | 3. Communal services | | |
| | | |   a. Education, including students' stipends and capital construction | 16.5 | |
| 4. Revenues of social insurance budget | | 6.6 |   b. Health care, including capital construction | 6.9 | |
| | | |   c. Total | | 23.4 |
| 5. Indirect taxes; other receipts from incomes of operating enterprises | | | 4. Government administration | | 4.4 |
|   a. Taxes on incomes of collective farms | .5 | | 5. NKVD | | 3.0 |
|   b. Revenues from profits of state enterprise | 9.3 | | 6. Defense | | 17.5 |
|   c. Taxes on incomes of cooperative enterprise | .8 | | 7. Financing the national economy | | 43.4 |
|   d. Turnover tax | 75.9 | | 8. Other | | 5.0 |
|   e. Other | 4.0 | | | | |
|   f. Total | | 90.5 | 9. Indicated budget surplus | | 3.1 |
| 6. Other | | 2.3 | 10. Total expenditures | | 107.0 |
| 7. Total revenue | | 107.0 | | | |

[a] For sources and methods, see Appendix D. Minor discrepancies between calculated sums of individual items and indicated totals are due to rounding.

On the expenditure side, the item "Financing the national economy," consists mainly of funds transferred to the banking system to finance the capital investments made in the economy. The investments ultimately take the form, in the case of those in fixed capital, of interest free grants and, in the case of those in working capital, of interest bearing loans to the economic organizations. The item "Financing the national economy" often is interpreted as consisting only of the foregoing kinds of outlays. Actually it comprises also various other outlays, principally stockpiling, subsidies to *khozraschët* organizations, appropriations for special measures of the sort referred to earlier, e.g., geological surveys. Finally, some funds turned over to economic organizations to finance workers' training programs may be included here rather than under the education item in the budget.

The residual expenditure item is a catch-all, comprising: (i) all expenditures of the social insurance system other than pensions and allowances; (ii) government outlays on communal services other than health care and education; and (iii) expenditures comparable to the items included in the residual item on the revenue side.

# 3 THE ADJUSTED FACTOR COST STANDARD OF NATIONAL INCOME VALUATION

The Adjusted Factor Cost Standard has the following features:

(i) All commodity prices resolve directly or indirectly into a series of charges for primary factors, particularly capital, land, and labor. Insofar as the prices may represent in part charges for materials, these resolve in turn into the charges for capital, land, and labor.

(ii) In the case of capital, there is a net charge and an "allowance for depreciation." The net charge, recorded either as a "cost" under the heading of "interest" or simply as a residual income, "profits," is at a uniform rate corresponding to the average productivity of capital in the economy generally. The average productivity of capital is understood to be an average of internal rates of return given by economies in costs realized through the introduction of capital in different branches. The allowance for depreciation supposedly is in accord with orthodox accounting principles.

(iii) The charge for land, "rent," corresponds on the average to the differential return to superior land.

(iv) The charge for labor, "wages," is at a uniform rate for any given occupation. At the same time, as between occupations, differences in wages correspond to the average difference in marginal productivities and also to differences in Disutility.[1]

(v) The principles in (iv) apply also to the relation of wages and farm incomes insofar as the latter represent the rewards of labor as distinct from capital or land.

(vi) Commodity prices are uniform within any market area.

I said I would comment here on the rationale and limitations of the Adjusted Factor Cost Standard as a norm for the measurement of "real" phenomena. This may best be brought out I believe by con-

[1] Here and elsewhere, Disutility is understood in the broadest sense, taking into account all factors influencing a worker's choice as between jobs, including not only arduousness, but also complexity, responsibility, etc. Those who are nevertheless averse to using this concept are free to think instead in terms of "competitive supply price," which is taken here to be its equivalent.

sidering the relation of this standard to certain ideal standards of national income valuation and certain applications of national income data which valuation in these terms permits. Let me first, then, explain briefly these ideal standards and applications. Readers familiar with recent theoretic writings on national income valuation will note that I do little more here than set forth in somewhat altered form the pertinent essentials of these studies.[2]

*Welfare Standard.* The different commodities entering into national income are valued proportionately to the Marginal Utilities of these commodities to consumers. In the case of investment goods, future returns are discounted at the prevailing Rate of Time Preference. In general the relative Marginal Utilities of different commodities might differ for different households; the Welfare Standard, however, presupposes an optimum disposition of consumers' goods among households in the familiar sense according to which this situation is excluded.[3]

Valuation in these terms, of course, is in order where the concern is to measure comparative Welfare. In principle, only if prices correspond to the Welfare Standard can national income data in "constant" prices be taken as they always are to measure differences in Welfare either for one country at different times or as between countries.[4] Although not usually thought of in the same connection, another familiar application, but one using data in "current" rather than "constant" prices, obviously also falls under this heading. I have in mind the case where data on the allocation of the national product as between alternative uses, e.g., consumption, investment, and the like, are taken as observations on the community's propensi-

---

[2] See especially J. R. Hicks, "The Valuation of Social Income," *Economica*, May, 1940; Simon Kuznets, "On the Valuation of Social Income," *Economica*, February, May, 1948; J. R. Hicks, "The Valuation of Social Income," *Economica*, August, 1948; P. A. Samuelson, "Evaluation of Real National Income," *Oxford Economic Papers*, January, 1950. I have also benefited from some privately circulated comments of Professor Samuelson on an earlier version of this study.

[3] Here and elsewhere where reference is to the question of the optimum resource allocation, the reader may find of value the brief survey in Abram Bergson, "Socialist Economics," in H. Ellis, ed., *A Survey of Contemporary Economics.*

[4] For a lucid exposition of the rationale of this application, see Hicks in *Economica*, May, 1940. Needless to say, there are also limitations, and on these reference is to be made again to this article of Hicks and also to the other writings cited above in note 2.

ties regarding income dispositions, e.g., the "propensity to save." [5]

By implication, I have referred here to a community in which "consumers' sovereignty" is a prevailing end and where accordingly Welfare is understood in terms of the Utilities of the individual households as they see them. In the present context it obviously is advisable to reckon also with the alternative case where resources are allocated in accord with a "collective preference scale" established by some political process or other, e.g., under socialism by decision of the Planning Board itself, and where accordingly Welfare is understood (at least by the Planning Board) in terms of these collective preferences and not necessarily in terms of consumers' Utilities. [6] In this case, of course, valuation is in terms of the planners' Marginal Rates of Substitution.

*Efficiency Standard.* The prices at which any pair of commodities is recorded in national income are inversely proportional to the relation of the Marginal Productivity of any given factor in the production of one of them to its Marginal Productivity in the production of the other. Insofar as the relative productivities of different factors vary as between industries, this principle leads to as many diverse results as there are factors. But according to familiar reasoning this awkward possibility is excluded if another optimum condition obtains, particularly the requirement that the community be operating on its Schedule of Alternative Production Possibilities, i.e., the alternative combinations of different products such that with the available resources the output of one cannot be increased without that of another being reduced. Valuation in terms of the Efficiency Standard presupposes this condition.

As appears only recently to have been clarified, [7] national income

---

[5] Insofar as the disposition of income in any particular period is supposed, as often is the case, to depend on "real" income, of course, this application too requires comparative data in "constant" prices. But although it is not always made explicit it generally is understood that the proportions in which income is distributed between different uses are given by data on the allocation of income in "current" rather than "constant" prices.

[6] In the case of communal consumption, of course, reference must be made in any case to a collective preference scale rather than to the utilities of individual households as such.

[7] Reference is to be made again to the writings of Hicks, Kuznets, and Samuelson.

data in "constant" prices measure not only comparative Welfare but also comparative Efficiency, understood in terms of shifts in the community's Schedule of Alternative Production Possibilities. But for this the pertinent valuation principle is not the Welfare Standard but the Efficiency Standard just described. A brief exposition is noted for a simple case.[8] Essentially, use is made of a schedule of Constant Dollar Output, i.e., for any given period the alternative combinations having in terms of prevailing prices a constant value. Given valuation in accord with the Efficiency Standard, this represents an upper limit to the community's production possibilities in the given period. And given this, one determines from the national income data in "constant" prices whether the actual output of one period is outside the range of the production possibilities of another.

I have spoken of the standard just described as *the* Efficiency Stand-

---

[8] Consider the simple community illustrated in the figure. The community produces only Butter and Guns, and in two periods considered in amounts designated by 0 and 1. The schedules of production possibilities for the two periods are shown as $Q_0$ and $Q_1$. Also shown are corresponding Schedules of Constant Dollar Output, $P_0P_0$ and $P_1P_1$. According to familiar reasoning, given that prices are in accord with the Efficiency Standard, these Schedules must be drawn as shown with the corresponding production 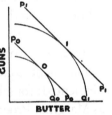 possibilities schedule either coinciding or lying to the left in each case. (Or rather this is so if there are Diminishing or Constant Returns; insofar as there are Increasing Returns, the production possibilities schedule might be to the right of the constant dollar output schedule. As the reader will readily see, the application is valid only in the former case.) From this it follows at once that alternative combinations of Butter and Guns which in terms of the prices of any given period have a greater dollar value than that actually produced at these prices, e.g., combination 1 compared with combination 0 at period 0 prices, necessarily lie beyond the production possibilities of the given period. Which is to say, of course, that a larger national income in one year in terms of the prices of another taken as base indicate the capacity to produce an output not previously feasible.

As will be seen, *decreases* in national income do not necessarily have a contrary implication. The reader will be aware that in this regard there is a parallel here to the application to Welfare, though interestingly as Samuelson has pointed out ("Evaluation of Real National Income," pp. 12 ff.), data which are decisive regarding changes in Efficiency cannot be at the same time decisive regarding changes in Welfare, and vice versa.

As with the measurement of Welfare, the measurement of Efficiency is subject to diverse limitations. One arising from Increasing Returns has been noted. On the limitations generally the reader may be referred to the theoretic writings already cited.

ard. Actually, Efficiency has been understood above in only one of several possible ways, and depending on the concept there are somewhat different standards. Insofar as it was required above that relative commodity prices correspond to factor Marginal Productivities for each and every factor, clearly reference was to Efficiency in a sense pertaining to the "long run" production possibilities schedule for which all factors, capital as well as labor and materials, are supposed to be allocable between different industries. Among the more interesting variants are these: (i) Reference is to Efficiency in terms of "short run" production possibilities, where labor and materials but not capital are supposed to be allocable between industries. In this case, the requirement concerning relative commodity prices may be somewhat relaxed. These prices must correspond to Marginal Productivity only for factors that are supposed to be allocable. (ii) Reference is to Efficiency in terms of "long run" production possibilities, but on the understanding not only that all factors are allocable but also that through retraining there may be shifts of workers as between occupations as well as between industries. In this case, the requirement concerning relative commodity prices has to be supplemented: for labor, commodity prices must reflect Disutility as well as Marginal Productivity.

As will be evident, there are corresponding differences in the conditions concerning resource allocation. In the case of the "short run" Efficiency Standard the community need be operating only on its "short run" production possibilities schedule while in the case of the two "long run" Efficiency Standards, the community must be operating on its "long run" schedule, with or without labor being shiftable between occupations, as the case may be.

The Efficiency Standard has been formulated so far in terms of factor Marginal Productivities and Disutilities. It is necessary to have in mind also another formulation. Given uniform factor prices, the correspondence of relative commodity prices to factor productivities evidently comes to the same thing as the equation of prices and Marginal Cost. At the same time, depending on what factors are supposed to be allocable, reference is either to "short run" Marginal Cost—which is to say Average Cost including rent to the "fixed factor"—or to "long run" Marginal Cost. Given constant returns to scale, the latter in turn comes to the same thing as "long run" Average Cost. In the case of such factors as are taken to be allocable, the factor prices in terms of which costs are calculated are supposed throughout to be proportional to factor Marginal Productivities in each and every industry. Finally, for the pertinent variant of the Efficiency Standard, wage differentials also correspond to Disutility.

In the case of the Welfare Standard, I referred to two kinds of applications, one the measurement of comparative Welfare, using global data in "constant" prices, the other concerning the community's propensities on income disposition, using data on the structure of the national product in "current" prices. Each of these has a counterpart in the case of the Efficiency Standard. On the one hand, given valuation according to this principle, national income data in "constant" prices, as has just been explained, provide a basis to measure comparative Efficiency. On the other hand, data in "current" prices on the allocation of the national product between investment, consumption, and the like provide a basis to calculate the alternative combinations of different kinds of output that the community is able to produce, e.g., taking the dollar value of the consumption sacrificed as an upper limit to the additional defense output made possible thereby, one calculates the community's defense production potential. This application is familiar, but it may not always be understood that it is in order only insofar as valuation is in accord with the Efficiency Standard. In the terms used here the application comes to the same thing as taking the Schedule of Constant Dollar Output as an upper limit of the Schedule of Alternative Production Possibilities. As was indicated earlier, this requires valuation in terms of the Efficiency Standard.

The theoretic standards of national income valuation have been formulated above in terms of rather abstract conditions on the price structure and without reference to the institutional arrangements under which the required price structure might be realized. For present purposes it is useful to have in mind also a few essentials on the latter aspect.

To refer first to a capitalist economy, the prevailing prices of course fully correspond at one and the same time to both the Welfare and the Efficiency Standard under "perfect competition" in the absence of distorting taxes. However, different features of "perfect competition" pertain to the different standards. For prices to correspond to the Welfare Standard it does not matter if there are sales taxes on consumers' goods or if monopoly power prevails in their production. But it would not do if there were consumers' goods rationing.[9] For prices to correspond to the Efficiency Standard, consumers' goods rationing is not necessarily of concern. But sales taxes and monopoly power at once cause deviations. Given a sales tax, for example, one can no longer suppose as is done in both the measurement of comparative Efficiency and the appraisal of alternative production possibilities that the dollar value of the Butter sacrificed is an upper limit to the dollar value of additional output of Guns made possible. If the tax is on Guns but not on Butter the increase in Guns may exceed the decrease in Butter, in dollar terms.

Strictly speaking the ideal is not "perfect competition" but the *equilibrium* of this system. Accordingly, one has to add to the list of causes of divergencies such aspects as windfall profits and rents which might occur under "perfect competition" itself. It will be evident, however, that in the case of the Efficiency Standard the situation in this regard differs depending on the variant.

For the socialist economy, one thinks at once of the familiar analogue of "perfect competition": the Competitive Solution of socialist planning.[10] As in the equilibrium of "perfect competition," so in the economic optimum of the Competitive Solution prices correspond completely to both standards.

The Competitive Solution, however, is only one of many possible planning schemes, and prices might possibly tend to correspond to the two standards under alternative arrangements. In the case of the Competitive Solution, the tendency of prices to correspond to these standards reflects the high degree of decentralization of decision-

[9] Actually a sales tax too appears to cause a deviation of prices from the Welfare Standard insofar as investment goods are valued in terms of discounted returns after taxes. In terms of consumers' Utilities the investment goods are undervalued in comparison with consumers' goods.

[10] See Abram Bergson, "Socialist Economics."

making, and together with this the fact that the Planning Board co-
ordinates decisions only indirectly through "trial and error" price
adjustments. Essentially, there is an open market for consumers'
goods and labor, while the managers of socialist firms are allowed
to determine for themselves their inputs and outputs subject to certain rules, i.e., cost economy and determination of the scale of output by the equation of Marginal Cost to price. The Planning Board
manipulates prices in order to assure that Supply and Demand correspond throughout the system.

It sometimes is argued that the Planning Board might approach
as closely if not more so to the optimum if in some measure it co-
ordinates input and output decisions directly while still manipulating
prices in some fashion to guide its own as well as its subordinate activities. The "trial and error" process is on paper rather than in the
"real" world.[11] No doubt critics of planning generally will find this
Centralist Scheme even more objectionable than the Competitive
Solution, but it would seem that the administrative short cut envisaged
is at least a logical possibility, and accordingly must be considered in
any complete account of the subject being discussed.

It was said that under capitalism a sales tax causes a divergence between prices and the Efficiency Standard. This is not necessarily so.
Conceivably the government might attempt through such a charge
to compensate for the undervaluation of some productive factor and,
insofar as this objective actually is achieved, evidently a correspondence of prices and the Efficiency Standard could be maintained. In
effect, a sales tax, understood in the conventional sense of a charge
based on the volume of sales and accruing as revenue to the government budget, does not represent as has been assumed a genuine addition to factor charges. Rather, it represents something of a factor
charge itself.

Of course, a situation of this sort in actuality must be rare under
capitalism, and a divergence between prices and the Efficiency Standard is a well-nigh universal result. On the other hand, such a situation seemingly might be of some importance under socialism, especially in a centralized type, where the government enjoys great

---

[11] *Ibid.*, pp. 410 ff.

discretion in deciding how to account for the different factors of production. Accordingly, while one might suppose that a divergence might be the usual result here too, the alternative is a real possibility. At the same time, what holds for the sales tax holds also for subsidies, understood as budgetary grants to cover accounting losses; and it is easy to see that there might be other elements in the socialist price system having different effects on the relation of prices to the theoretic standards, depending on the circumstances.

Actually, as will appear, these theoretic considerations have some application in the Soviet Union. On the other hand, it is only fair to say now that our appraisal of ruble prices in the chapter following attempts to proceed without any detailed inquiry into this aspect.[12]

To come finally to the relation of the Adjusted Factor Cost Standard to the theoretic standards, the essentials can now be readily set forth. If the reader will refer again to the definition of the Adjusted Factor Cost Standard he will see at once that this comes to the same thing as Average Cost, understood as not necessarily the cost actually recorded. Rather it is what would be recorded if factor prices were uniform as between industries and at the same time corresponded to

[12] It may be in order to comment briefly here on some special aspects of the problem of valuation discussed in the theoretic writings to which reference has been made at various points, particularly those of Hicks and Kuznets. First, while I am committed in this study to treat all of government as a final product (see above, Chapter 2, section on methodology), there is no reason to dissent here from the view of Hicks and Kuznets that in principle some of this category should be treated as intermediate. As to whether this is also a practicable procedure, as Kuznets affirms ("On the Valuation of Social Income," pp. 6 ff.) and Hicks has conceded (in his second essay in *Economica*, 1948, pp. 164 ff.), is a question which it may be just as well to pass by. Second, given direct taxes and free government services that affect factor supply prices, there seems to the writer to be no alternative but to recognize that there is no fully valid Efficiency Standard where the concern is with "long run" production possibilities taking into account workers' shifts between occupations as well as industries. Relative prices that correspond to factor productivities cannot at the same time correspond to Disutilities. This is to be contrasted with Kuznets' view (p. 123), apparently accepted by Hicks (pp. 165 ff.), that the valid Efficiency Standard in these circumstances is one for which factor incomes are included net of direct taxes and gross of free public services affecting factor supply prices. Third, the writer agrees with Hicks (p. 168) that contrary to the implied view of Kuznets (p. 123) free public services cannot be supposed generally to affect factor supply prices. Fourth and last, it seems necessary to follow Hicks in rejecting Kuznets' attempt to reconcile the Welfare and Efficiency Standards. Kuznets relies in this connection on the views to which the writer has just taken exception.

relative factor productivities *on the average* in the economy generally. In the case of labor there is an analogous relation of wages to Disutility.[13]

Suppose now the community is on its production possibilities schedule, and more particularly the "long run" one where there are shifts of workers between occupations. In this case, relative factor productivities are the same in each and every industry. Accordingly, factor prices that correspond to relative factor productivities *on the average* also correspond to these productivities in each industry. Also, there is a similar extension of the relation of wages to Disutility. This is to say, then, that the Adjusted Factor Cost Standard comes to the same thing as the Efficiency Standard, given operation on the production possibilities schedule. The production possibilities schedule is the "long run" one just referred to, and reference is particularly to the variant of the Efficiency Standard pertaining to this schedule.[14] But, of course, if the Adjusted Factor Cost Standard corresponds to this variant it would also correspond to the others, which are less restrictive.

Suppose in addition the consumers' goods are disposed of among households in an optimum manner, and furthermore there is an optimum "bill of goods" in the familiar sense, in terms of consumers' Utilities or planners' preferences as the case may be, i.e., if reference is to consumers' Utilities, the ratio of the Marginal Utilities of any two commodities corresponds to the rate at which one can be substituted for another through resource reallocation. In this case, it is self-

---

[13] In the case of materials which are at the same time a factor and a product, an awkward possibility is that a price that corresponded on the average to the relative productivity of this article in comparison with other factors would not at the same time correspond to Average Cost in terms of the given prices of factors in general. It is a familiar fact, however, that there are generally relatively limited possibilities of substitution between materials and other factors so one may be justified in assuming that any of a wide range of prices might correspond equally well to relative factor productivities in the case of materials. On the other hand, where this is not the case, the Adjusted Factor Cost Standard and Average Cost as understood are not the same thing. I believe I am right in thinking that in the case of capital goods where there is an additional degree of freedom to determine the value of the asset there is no comparable problem.

[14] I am assuming of course constant returns to scale. Otherwise the Adjusted Factor Cost Standard still diverges from the Efficiency Standard to the extent of the divergence between "long run" Average and Marginal Cost.

evident that the Adjusted Factor Cost Standard corresponds also to the Welfare Standard.

On the other hand, if the bill of goods and their disposition are not optimal this is not the case. Furthermore, if the community is not on its "long run" production possibilities schedule the Adjusted Factor Cost Standard also diverges from the Efficiency Standard. Of course the community might still be on its "short run" production possibilities schedule but, even so, the Adjusted Factor Cost Standard diverges from the variant of the Efficiency Standard pertaining to this schedule. While valuation at Adjusted Factor Cost allows only an average return on capital, this variant of the Efficiency Standard requires the inclusion in prices of the entire "short run" profit to capital.

In the "real" world, of course, the optimum conditions are never fully met, so in actuality the Adjusted Factor Cost Standard never could correspond to either of the two theoretic standards. But clearly depending on the efficiency of resource allocation it approximates them more or less closely. Let me explain, then, that it is as such an approximation that the Adjusted Factor Cost Standard has been selected for study here. Given the inefficiency, the theoretic standards are themselves unattainable. There is no valuation for which national income data are theoretically valid measures of the "real" phenomena described. It may be hoped that with valuation at Adjusted Factor Cost the range of conjecture is less than it might be otherwise.[15]

But what of resource allocation in the USSR? To what extent is

[15] As is mentioned below and in any case is surely almost inevitable, resource allocation in the Soviet Union as in Western countries tends to approach more closely "short run" than "long run" production possibilities. Under the circumstances, the Adjusted Factor Cost Standard presumably approximates more nearly to the "long run" than to the "short run" variant of the Efficiency Standard. The approximate in neither case, however, could be especially close, and the possibility is suggested that, as an alternative, attention might be focused on an Adjusted Factor Cost Standard which would approximate more nearly the "short run" variant. Thus, an attempt would be made to allow for "short run" profits to capital rather than, as I have, merely to include an average return. Given the closer approach of resource allocation to "short run" production possibilities it might be hoped that the end result while still only an approximation would be more nearly valid theoretically. This alternative, I believe, is indeed an attractive one in principle; if I have not endeavored to apply it here it is simply because of the limitations of data which hardly permit a calculation of "short run" profits.

Of course, in Western countries, the tendency of resource allocation to approximate

the optimum realized? From what is said in the chapter following on the relation of ruble values to Real Costs and from generally known facts, the reader will have some impression of Soviet economic efficiency. But it may be hoped that it will be understood, too, that no attempt can be made here to deal explicitly with this large question. Although it obviously is a major concern, the extent to which valuation at Adjusted Factor Cost approximates the ideal has to be left an open question here.

But obviously even in the best of circumstances the approximation could not be especially close; and it still remains to be seen to what extent the Adjusted Factor Cost Standard itself is realizable with available statistical data. Why not, then, do as Clark and Wyler do after all; that is, abandon the ruble standard altogether, and use instead non-Soviet prices? The writer does not deny that this alternative is tempting. But before renouncing calculations in terms of rubles once and for all, the reader should ponder, too, the limitations of non-Soviet prices.

Both Clark and Wyler resort to United States dollar prices: the former, the average dollar prices of the period 1925-34 (the calculations are in terms of "international units" having the average purchasing power of the dollar in this period); and the latter, dollar prices of 1940. But the dollar prices necessarily reflect the preferences, technology, and cost structure, not of the USSR at the time studied, but of the United States. Accordingly, the national income data might be used in comparisons of Welfare or Efficiency as between the United States and Russia, though obviously even here there is only a partial basis for an appraisal. But, except by chance, they could hardly be used to appraise the prevailing Soviet propensities on income disposition or alternative production possibilities; or for intertemporal

---

more to "short run" than to "long run" production possibilities is associated with a corresponding tendency of prices to approximate more to the "short run" than to the "long run" variant of the Efficiency Standard. Given this there is at once an impelling case for according priority to the former variant; and this is the general practice insofar as use usually is made of prevailing prices without any revaluation. But while there is a parallel to Western experience in the case of resource allocation there unfortunately is little basis to think there is one in the case of prices as well, so any arguments for the "short run" variant from this standpoint would be unfounded. As to this view on Soviet price formation, the information set forth in the next chapter may be illuminating but for present purposes it probably is best to consider it as a working hypothesis.

comparisons of Welfare or Efficiency for the USSR alone. For such purposes, the calculations almost inevitably are distorted. Thus, suppose as seems true that the United States technical superiority is more pronounced for highly fabricated than for little fabricated goods. One might expect United States dollar prices for investment goods, then, to be relatively low in comparison with consumers' goods. The reduction of investment in the USSR by one dollar in terms of United States prices might actually release resources sufficient to produce much more than a dollar's worth of consumers' goods in terms of the same prices, rather than less as is supposed in appraising alternative production possibilities.

Furthermore, it is self-evident that any limitations in our calculations arising from deviations of Soviet resource allocation from the optimum must inhere also in calculations in terms of dollar prices.

Finally the dollar price structure is something less than ideal even in reference to the appraisal of "real" phenomena in the United States. I have in mind, of course, such aspects as monopoly power, taxes and subsidies, and regulated prices.[16]

[16] Without attempting to outline an alternative, the writer has commented elsewhere on the limitations of the dollar valuation procedure. See Abram Bergson, "Comments" in "The Economy of the USSR," Papers and Proceedings of the Fifty-ninth Annual Meeting of the American Economic Association, *American Economic Review*, 1947, No. 2, pp. 643 ff. Maurice Dobb also discusses the problem in *Soviet Economy and the War*, pp. 37 ff.

The reader familiar with the earlier version of this study in the *Quarterly Journal of Economics* for May and August, 1950 ("Soviet National Income and Product in 1937, Part I: National Economic Accounts in Current Rubles; Part II: Ruble Prices and the Valuation Problem") will have noted that my treatment of the valuation question is now appreciably altered. Among other things, the Adjusted Factor Cost Standard as now understood represents a combination of a "real" cost standard, previously defined, together with another so-called "physical volume of resources used" standard that also was considered previously. The precise nature of and basis for this revision I believe are sufficiently evident, but it should be noted that the revision is intended in part to take account of illuminating comments on the earlier version in memoranda prepared for private circulation by Professor Paul A. Samuelson and Mr. Norman Kaplan. Also, I have taken the occasion now to discuss more explicitly than before the rationale of the standard adopted and in this connection acknowledgment is due again to these memoranda, especially Professor Samuelson's. Finally, mention should be made of a third memorandum by Professor Jacob Marschak. Among other things, this sets forth the interesting proposal that in view of the limited substitutability as between major categories of goods, e.g., food, plant construction, munitions, aggregation should be limited to summations within these categories and should not extend to summations of the categories themselves. The limited substitutability is from the standpoint of war potential.

# 4 NATIONAL INCOME AT ADJUSTED FACTOR COST

### INTRODUCTION

To turn now to the relation of ruble prices to Adjusted Factor Cost and the effects of divergencies on our ruble accounts, which is the subject assigned this chapter, it may be advisable first to recall the essential features of this standard: (i) All commodity prices resolve fully into charges for primary factors, particularly capital, land, and labor. (ii) For capital, there is a net charge, corresponding to the average internal return on this factor in the economy generally, and an allowance for depreciation of a conventional sort. (iii) The charge for land, "rent," corresponds on the average to the differential return to superior land. (iv) "Wages" are at a uniform rate for any occupation and as between occupations differ on the average in accord with differences in productivity and Disutility. (v) Similar principles apply in the case of the relation of wages to farm labor income. (vi) Commodity prices are uniform in any given market area.

Apparently the relation of retail prices to consumers' Utilities is not of immediate concern here, but it has to be considered nevertheless since it affects the meaning of money wages and also, as will appear, is of interest in certain other connections.

The discussion proceeds as follows: In the section immediately following, I consider in turn different features of the ruble price system other than retail prices, wages, and farm household income, and ask in each case how the relation of ruble prices to Adjusted Factor Cost is affected. Reference is primarily to the circumstances prevailing around 1937, the year to which the national income data relate. The same sort of inquiry is made concerning retail prices and wages in the section on this subject (pp. 63–68), and concerning farm household income in the section on collective farm incomes (pp. 68–74). Later (pp. 74–85), I appraise and, where feasible, correct for distortions in our ruble accounts due to such divergencies as are disclosed. In this connection, I focus exclusively on two tabulations, those of the national income by use and by industrial origin, these being the ones which I believe are of chief interest for the purpose in hand. A final section (pp. 86 ff.) is devoted to a comparison of the revised tabulation of the national income by use with corresponding ones derived in dollar terms by Clark and Wyler. The comparison may be of interest from various standpoints.

#### RUBLE PRICES VERSUS ADJUSTED FACTOR COST: GENERAL

*The turnover tax.* For purposes of this discussion I shall understand a "sales tax" as a charge against sales revenue based on the volume of sales and yielding a corresponding income to the government budget. The famous turnover tax of the USSR probably is a far more complex fiscal device than is commonly supposed, but in general it clearly falls within the terms of the definition of a sales tax just given. While levied in a variety of ways the amount of the tax varies with the volume of sales. Thus the tax is levied most often as a flat percentage of selling price, but it also takes such diverse forms as a fixed absolute charge per unit of product, the difference between the wholesale and retail price after allowing for a standard trading margin, and so on.[1]

It was pointed out in the previous chapter that a sales tax under socialism might possibly represent a factor charge rather than any actual addition to such charges. In the Soviet Union, this may actually be the case in regard to the taxes levied on procurement agencies. Details on this will be set forth subsequently. Possibly there are also other instances. But I believe it is not especially important here to make any exhaustive study of this aspect; in any event, in trying later on to appraise the distortions in our ruble accounts, I propose to assume provisionally that the turnover tax is not in any part a factor charge. Insofar as it is indeed such a charge, this presumably must be reflected in an undervaluation in the factor in question, and this can be considered separately.

Insofar as the tax is not a factor charge, of course, there is no place for it in our Adjusted Factor Cost Standard. Accordingly, the turnover tax must be reckoned as an initial cause of divergence between ruble prices and Adjusted Factor Cost.

[1] On the turnover tax generally, I have found most helpful the detailed discussion in A. K. Suchkov, *Dokhody gosudarstvennogo biudzheta SSSR.*

At the same time, such a divergence evidently is of concern here only insofar as it varies as between the different elements of our national income tabulations. But for reasons revolving around the fact that some goods, intermediate products, are used to produce other goods, even a uniform sales tax on all commodities would have this effect.[2] Also, a notable feature of the turnover tax is the wide variation in rates. The variation is greatest as between heavy industrial and consumers' goods, but there is also considerable differentiation within these categories. Thus, as of 1937, the tax on coal and steel was .5 percent; and on most types of industrial machinery, 1.0 percent. At the same time, the tax on men's leather shoes, depending on the make, varied from 17 to 35 percent. For most types of cotton textiles, the tax was 44 percent and appreciably higher.[3] Under the circumstances, it turns out that the turnover tax is not only a cause of divergence between ruble prices and Adjusted Factor Cost but by far the most important one with which we have to deal.

*Subsidies.* Of a piece with the sales tax, although with opposite effects on the relation of prices and Adjusted Factor Cost, is a subsidy which permits operation at a price less than that corresponding to the charges for productive services. As is already clear this, too, is a feature in the USSR. While no longer as important in industry in 1937 as they had been before the price reform of April, 1936, subsidies still were paid in many basic industries. Also, there were payments in other sectors, particularly the MTS, to compensate for undervaluation of income in kind received from the collective farm in payment for services rendered. Further details on this element in prices are presented later. Presumably one should have in mind here a possible situation similar to that just referred to regarding the turnover tax: conceivably the subsidy might compensate for an overvaluation of some factor. But as with the turnover tax, I shall pass by this

[2] It should be noted that the question of the effect of an indirect tax on relation of prices and Adjusted Factor Cost, which is discussed here, is similar to but not quite the same as that of the effect of an indirect tax on resource allocation, as discussed, for example, in Abram Bergson, "Socialist Economics."

[3] See Appendix E, pp. 135 ff. The cited percentages are based on the wholesale price gross of the tax.

complexity in the appraisal of distortions in our ruble accounts, so the subsidies will be reckoned *in toto* as still another cause of divergence between ruble prices and Adjusted Factor Cost.

*Interest and profits; depreciation.* Except in the case of short-term loans obtained from the State Bank for seasonal and other special working capital needs, no interest is charged for capital in the USSR. Rather the operating enterprise obtains its capital in the form of interest free grants.[4] On the other hand, an interesting feature is the inclusion in prices of a planned profit. Except in the case of subsidized enterprises prices generally are fixed to allow the enterprise a profit for operation in accord with the plan. Furthermore, there necessarily are unplanned profits and losses due to deviations from the plan.

For purposes of this study, I assume provisionally that the planned profit is an arbitrary figure, and in any event not a *bona fide* charge for capital. Regrettably Soviet sources have little to say on the principles according to which planned profits are fixed and, while one might possibly wish to examine in this connection what role if any this category plays in Soviet planning, I believe the question of whether it might stand for the uniform charge on capital called for in the Adjusted Factor Cost Standard may just as well be appraised statistically as I propose to do subsequently. Of course here, as in the case of the turnover tax and subsidies, one has to be aware that under socialism a given charge may mean different things, and the possibility is open that the planned profit might actually be a charge for some factor other than capital. But I believe the first hypothesis stated is the only one worth exploring here, and in any event the reader will see that the end result of the analysis is hardly affected by our preoccupation with this aspect. As to the extra-plan profits presumably these are to be classed either as wages of management, insofar as they represent superior administrative ability, or alternatively as the counterpart of the windfall gains of capitalism.

---

[4] As under capitalism so in the USSR, the interest on short-term loans may be considered as the price charge for banking services. Whether and to what extent this particular price corresponds to Adjusted Factor Cost is not a question which can be explored here, but it should be noted that the State Bank of the USSR, which is responsible for the short-term loans, apparently operates on the same *khozraschët* principles as economic enterprises generally.

In this connection, then, one must reckon with two possible sources of divergencies between prices and Adjusted Factor Cost: one due to the omission from the costs of the enterprise of a charge for capital; the other due to the inclusion in prices, under the heading of "profits," of a charge which only in part represents a productive service rendered. In this connection the reader will be aware that even windfall profits represent a divergence from Adjusted Factor Cost.[5]

As under capitalism so in the USSR, costs include charges for depreciation. Whether and to what extent these charges, as often is suggested, overestimate the service life of fixed assets under conditions of the five year plans is a question which must be left to an independent inquiry. The charges must be on the low side in any event, insofar as they are calculated in terms of original cost. In view of the price inflation under the five year plans, the corresponding allowances in terms of replacement cost undoubtedly would be much higher.

*Agricultural rent.* For reasons to appear, it is necessary now to elaborate somewhat the brief comment in Chapter 1 on the complex procurement system of the USSR. A part of the produce is taken by the government simply as a payment in kind for the services rendered the collective farms by the machine-tractor stations. Another part is taken by the government in the form of obligatory quotas at quite low prices. Still another part is procured on a voluntary basis at appreciably higher prices. And finally, farmers to whom urban collective markets are accessible, are also able to sell their surpluses at still higher prices there.[6]

Data on the government procurement as distinct from collective

[5] They also represent a divergence from the Efficiency Standard pertaining to "long run" production possibilities, but not from that pertaining to "short run" production possibilities.

[6] Because of the difference in prices as between voluntary sales to the government and sales on the collective farm market, a question is sometimes raised as to whether the former are in fact voluntary as the Russians say. Perhaps the answer is in part that they are not entirely so, but it should be observed that there are factors which might induce the farmer to sell voluntarily to the government rather than on the collective farm market, even though the government prices are lower than those on the collective farm market. One such factor is that, as is implied in the text, the urban market is not accessible to many farmers in distant localities. Another is that the government at various times has made a practice of granting to the farmer making such sales priorities for the reciprocal delivery of scarce manufactured goods.

farm market prices are sparse but some notion of their general level
is readily obtained from data on the turnover tax. The aggregate value
of government procurements of farm products in 1937 amounted to
15.7 billion rubles at realized farm prices.[7] According to the projected
budget for 1937, procurement agencies were to pay 24.1 billion rubles
in turnover taxes on their procurements, while the food and light
industries were to pay another 31.7 billion rubles in turnover taxes
on their own sales.[8] The final retail selling prices of the products of
these industries necessarily cover not only all procurement, processing, distribution, and transportation costs but also these turnover
taxes. Evidently the total sum paid to farmers for their produce must
have constituted only a small fraction of the retail selling price, exclusive of procurement, processing, distribution, and transportation
costs. The same general conclusion is indicated by other information
at hand.[9]

According to L. E. Hubbard [10] the average price paid by the gov-

[7] Gosudarstvennaia Planovaia Komissiia (State Planning Commission), hereafter
abbreviated as Gosplan, *Tretii piatiletnii plan razvitiia narodnogo khoziaistva Soiuza
SSR, 1938–1942* (Third Five Year Plan of Development of the National Economy of
the USSR, 1938–1942), p. 88. Presumably the cited figure for procurements includes
the payment in kind for the services of the machine-tractor stations.

[8] See Appendix E.

[9] (i) As of September, 1935, the government procurement agency's own wholesale
price of grain to millers (freight prepaid to the railway station of destination) in
price region III (which includes the Moscow *oblast'*) was 93 rubles per centner (100
kilograms). See G. I. Kuznetsov, editor, *Sbornik otpusknykh i roznichnykh tsen i
torgovykh nakidok na prodovol'stvennye tovary* (Handbook of Wholesale and Retail
Prices and Trading Margins on Food Products), p. 19. Out of this sum, the procurement agency had to pay a tax to the government, a turnover tax amounting to 80
rubles per centner. See V. Petrov and V. Fisherov, compilers, *Nalog s oborota* (The
Turnover Tax), p. 4. The balance of 13 rubles per centner or 14 percent of the price to
millers presumably represented in part the procurement agency's own costs and in
part the net receipts of the farmer from grain deliveries. It should be noted that in
the case of bread the turnover tax was levied entirely at the procurement stage.

(ii) L. E. Hubbard (*The Economics of Soviet Agriculture*, p. 215) estimates that
the ratio between "the cost to the government of grain and the retail price of the same
weight of bread amounted in 1937 to about 1:7 for rye and 1:8 for wheat."

(iii) According to S. N. Prokopovicz (*Quarterly Bulletin of Soviet Russian Economics*, December, 1940, p. 76) the government's price for obligatory deliveries in the
Altai province as of 1940 was 2.40 rubles a kilogram in the case of meat, and .45 rubles
a liter in the case of milk. At the same time, the prevailing prices for these products
in the retail collective farm market were respectively 12–15 rubles a kilogram and 2
rubles a liter.                                    [10] *Op. cit.,* p. 215.

ernment for voluntary deliveries of grain in 1937 was 20 to 30 percent above the prices for obligatory deliveries, and for large quantities there were premiums running up to the amount of the base price itself.

The government's quotas for obligatory deliveries of produce are not the same throughout the USSR but vary regionally.[11] Apparently these differences in quotas correspond to differences in fertility; it is believed that the quotas also vary in dependence on the distance from the urban market center.

Charges in kind for the services of the machine-tractor stations similarly vary not only with the amount and kind of service rendered but also with the harvest yield. For example, for spring plowing of grain fields, the charge per hectare varied from 9 kilograms, in the case of farms with a yield of less than 300 kilograms per hectare, to 130 kilograms in the case of farms with a yield of more than 1,300 kilograms per hectare.[12]

Finally, while procurement prices vary as between obligatory and voluntary deliveries, this difference generally does not accrue to the processor. For obvious reasons, the government policy is that processors should be on a more or less equal footing in regard to raw material costs, regardless of the manner of procurement. Thus differences in procurement prices generally are in large measure offset either by corresponding differences in the turnover tax or by refunds from turnover tax funds to purchasers of produce procured on a voluntary basis or apparently in some cases by subsidies to purchasers of such produce.[13]

The point towards which the foregoing remarks are aimed should be evident. To some limited extent the turnover tax on agricultural produce is to be regarded not as a sales tax but as the economic counter-

[11] *Ibid.*, p. 186.          [12] *Ibid.*, p. 151.

[13] See above, p. 25, note 5. In the case of agricultural produce, then, there are in effect multiple turnover tax rates for individual commodities. Such multiple rates also prevail in other cases, but these do not seem to hold the same interest for us here as that of agricultural produce. Thus there are sometimes differential rates as between economic organizations (e.g., producers' cooperatives may be allowed lower rates than state enterprise; Labor Reserve School production is freed altogether from the tax, and so on) and also as between regions (e.g., sales in distant regions may be subject to a lower tax or exempted altogether).

part of agricultural rent. As a result of variable obligatory delivery quotas and low prices for these deliveries, the farmer with better land is in effect being charged a rent for this land.[14]

It remains to observe that the government does not extract in this way all of the differential income produced by superior land. Some part of this income remains in the hands of the farmers. Subject at any rate to the proviso that difference in capital investment must be a factor, this appears to be the implication to draw from the substantial variation in farm incomes. While in 1937 the collective farms of the USSR on the average distributed 4 kilograms of grain per labor day to the member households, over 17 thousand farms distributed from 7 to 10 kilograms per labor day, and 126 farms distributed more than 20 kilograms per labor day.[15]

I need hardly add that land in the USSR is *formally* rent free. While the land in principle belongs to the government, no charge for rent is made under this name, a fact which Soviet writers do not hesitate to cite as an advantage of their system.[16]

*Multiple prices.* The correspondence of prices and Adjusted Factor Cost, it has been stated, reflects in part the fact that there is within any given market area a uniform price for any one commodity. What seems to be the outstanding case where this principle is violated in the USSR has already been referred to. This is the case of agricultural procurement prices. But reference here is to raw materials and as was indicated, through turnover taxes and subsidies, the prices charged the processors of these materials are made more or less uniform. Accordingly, so far as the prices of processed goods are

[14] Insofar as there is an undervaluation of agriculture due to subsidies, the turnover tax in part might be viewed also as an offset to this aspect.

[15] TSUNKHU, "Sotsialisticheskoe sel'skoe khoziaistvo Soiuza SSR" (Socialist Agriculture in the USSR), reprinted in *Planovoe khoziaistvo* (Planned Economy), 1939, No. 7, p. 144.

[16] Soviet economists, however, acknowledge the existence of economic rent in the USSR, insofar as reference is to the differentiation in agricultural incomes that result from differences in land fertility and location. A recent article on this subject also acknowledges, though rather obliquely, that in effect a part of this differential income accrues to the government rather than to the farmers. It is maintained, however, that "absolute rent," in the Marxian sense, has been abolished. See I. Laptev, "Kolkhoznye dokhody i diferentsial'naia renta" (Collective Farm Income and Differential Rent), *Bol'shevik*, August, 1944, pp. 8 ff. A translation appears in the *American Review on the Soviet Union*, May, 1945.

concerned, there is no major divergence of prices and Adjusted Factor Cost at this point.

At the same time, the use of taxes and subsidies to establish uniform prices for agricultural produce, while a general rule, does not appear to have been universal. In the case of grain at least, supplies are believed to have been made available both for stockpiles and for export purposes at the low prices the government allowed the farmers for obligatory deliveries.[17]

A question arises also concerning the relation of retail collective farm market and retail state shop prices. But while these have often diverged widely, there is some basis to think that at the time considered in this study they must have been more or less in line. The reasons for this view are set forth in the following section.

It seems advisable to leave to a separate inquiry the question of whether and to what extent the armed forces might be favored by the ruble price system. The writer has been unable to find any evidence in Soviet sources to support occasional reports that prices are differentiated in favor of military procurements, either through special exemptions from the turnover tax or otherwise; and some information is to the contrary. But for the present the question probably should be considered an open one.

### RETAIL PRICES AND WAGES

A correspondence of retail prices to consumers' Utilities might conceivably prevail under diverse circumstances, but in any actual case it is more than doubtful that it could be approached at all unless there were an open market for consumers' goods. Furthermore the correspondence would depend on the effectiveness of the market, i.e., the degree to which at prevailing prices available supplies suffice to meet demands. In the USSR, as was mentioned in Chapter 1, rationing rather than the open market was the rule from the beginning of the five year plans until 1935 and 1936 and again after the Nazi attack.

At the same time, the indications are that the open market that

[17] Kutler, *Gosudarstvennye dokhody SSSR* (Government Incomes of the USSR), p. 88.

prevailed in the interval between these periods operated with varying effectiveness. On this, the data in Table 7 are illuminating.

## TABLE 7

### RELATION OF COLLECTIVE FARM MARKET AND STATE RETAIL SHOP PRICES, SPECIFIED COMMODITIES

(*State retail shop price = 100 percent*)

| COMMODITY | 1934, LOCALITY NOT SPECIFIED IN SOURCES [a] | DECEMBER 25, 1935, MOSCOW [b] | JANUARY 1, 1940, MOSCOW [c] |
|---|---|---|---|
| 1. Beef | 233 [d] | 87 | 143 [e] |
| 2. Pork | .. | 81 | .. |
| 3. Potatoes | 923 | 137 | 1200 |
| 4. Cabbage | .. | 167 | .. |
| 5. Onions | .. | .. | 167 |
| 6. Beets | .. | .. | 300 |
| 7. Carrots | .. | .. | 400 |
| 8. Milk | 343 | 129 | 238 |
| 9. Eggs | 205 | 97 | 141 |
| 10. Butter | 220 | 87 [f] | .. |
| 11. Average | 384 | 112 | 370 |

[a] L. E. Hubbard, *Soviet Trade and Distribution*, p. 148.
[b] TSUNKHU, *Kolkhoznaia i individual'no-krestianskaia torgovlia* (Collective Farm and Independent Peasant Trade), pp. 178 ff.
[c] S. N. Prokopovicz, *Quarterly Bulletin of Soviet Russian Economics*, May, 1941, p. 128.
[d] "Meat."
[e] "Beefsteak."
[f] Melted (i.e., preserved) butter.

As the one free market in the USSR, the collective farm market
inevitably bears the brunt of any shortages in supplies in state retail
shops; a shortage in supplies available at established prices in the
state shops necessarily means high prices in the collective farm market. This in a word was the situation that prevailed in 1934, when
rationing still was in effect and only restricted rations were available at state shop prices. It was also the situation that prevailed in
1940 when, as a result of the increased military procurements and
stockpiling and the continued expansion of employment, there was
again a general shortage in the state shops.

In December, 1935, however, the rationing of foodstuffs had just been liquidated and, as a preparation for this step, state shop prices had been raised much above the level under rationing. At these prices, and with the supplies then available, effective demand and supply presumably were in reasonably close correspondence, so there was in general no need for the consumer to pay a large premium for supplies in the collective farm market.[18]

Unfortunately, data on the relation of collective farm market and state retail shop prices in 1937 are not at hand for this year. But it appears that retail prices in state shops, at least for foodstuffs, were fairly stable from the latter part of 1935 through 1938,[19] and in view of what is generally known about the economic conditions of 1937, it is reasonable to assume that if anything the consumers' goods market operated with greater effectiveness in this year than in December, 1935. While employment and purchasing power had expanded further, so also had available supplies: as has been mentioned, 1937 was for the Soviet consumer a year of relative prosperity.[20]

For wage differentials to tend to correspond to productivity and Disutility under socialism, seemingly the following features must prevail:

(i) The "wages" in question represent at one and the same time

[18] The fact that in some instances the prices in the collective farm market were actually below those in the state shops is rather puzzling when it is considered that the state shop patron was always free to buy in the collective farm market. The explanation presumably is to be found in market imperfections, such as comparative convenience of location of the farm markets and state shops, and perhaps in differences in the quality of goods. Data available for other cities for 1935 indicate that the relations prevailing in Moscow were reasonably representative at that time.

[19] According to calculations of S. N. Prokopovicz (*Quarterly Bulletin of Soviet Russian Economics*, November, 1939, p. 55; May, 1941, p. 130), the average level of food prices in 1937 was only a few percent and in 1938 about ten percent higher than in October, 1935. While Prokopovicz is not explicit it seems clear from the context at least in the case of the second figure that reference is to state shop rather than collective farm market prices.

[20] To this general evidence on the state of the Soviet retail market in 1937, it may be in order to add a comment based on the writer's own first-hand observations during a visit to the USSR in the summer of 1937. During this period, the writer visited a great many state retail shops mainly in Moscow, and can report that while there was some queuing up in some of them, particularly in the case of meat stores, the general impression gained was one of orderly marketing, with stocks generally adequate to meet requirements at the established prices.

the charge for labor recorded as costs for the employing enterprise and incomes paid out to the worker.

(ii) The employing enterprise or superior agencies seek systematically to economize money costs.

(iii) There is an open market for labor in the sense that the worker is free to choose his occupation at the prevailing wage rates.

(iv) The labor market operates effectively in the sense that wage differentials are more or less responsive to divergencies between the "demand" and "supply" of labor in the different occupations, as determined by the foregoing circumstances.

(v) There is likewise an open market for consumers' goods and this too operates effectively in the sense that at prevailing prices there are sufficient goods to meet the demand.

Of course, not all features need prevail for wages to correspond to either productivity or Disutility alone. Rationing, for example, tends to cause a divergence between wages and Disutility but need not affect the relation of wages and productivity.

To what extent do the foregoing features prevail in the USSR? It already has been made clear that those concerning consumers' goods market do so only to a limited degree, at least if account is taken of the lengthy period that obviously is relevant to workers' occupational choices and not merely of the year 1937; and this presumably means a divergence between wages and Disutility. But obviously it would be easy to put undue stress on this aspect; in any event the writer ventures to think that, if account is taken of other available information bearing on the relation of wages and Disutility and also productivity, there may be sufficient correspondence in both regards for the purposes of this study; accordingly, no divergence between ruble prices and Adjusted Factor Cost will be reckoned with at this point. The other information is set forth below:

(i) Reference already has been made to the *khozraschët* system in the USSR, according to which separate records are kept in money terms of the charges for different factors and the proceeds from sales.[21] Under this system one of the charges is that for labor. Fur-

---

[21] Reference may be made to David Granick, *Plant Management in the Soviet Industrial System,* doctoral dissertation, Columbia University, 1951, particularly Chs. IX,

thermore, an established complement to the *khozraschët* system is that the individual enterprise as well as superior agencies are expected to concern themselves with cost economy. At the level of the individual enterprise the government seeks to sharpen this concern by paying premiums to managerial staff which are calculated on the basis of cost economies achieved; at all levels, cost reduction is one of the bases for appraising success.

A complicating feature is the fact that at least up to the war financial controls were still rather lax. Under the circumstances, enterprise managers often have tended to subordinate cost consideration in the interest of fulfilling and overfulfilling high-priority production targets.[22] On the other hand, insofar as the financial controls are effective the concern for cost economy is reinforced by the concern to fulfill and overfulfill production targets.[23]

(ii) The wages charged as costs in the USSR are paid out to the workers as income. This is subject to the proviso that there is an income tax, which until the war was for the bulk of the workers very limited in magnitude and only mildly progressive.[24] Also, as already has been mentioned, there was an open market for labor in the USSR during the entire period from war communism to the year 1940. While there were some controls, these apparently were of relatively limited importance; in large measure the Soviet worker was free during this period to choose for himself his occupation.

(iii) In contrast to the equalitarian rationing practiced in the United States in wartime, Soviet rationing was characterized by extensive differentiation, with preferential treatment being accorded to the more important industries, plants, and also the more productive workers. Also, even during the rationing period there was an

---

X; G. Bienstock *et al., Management in Russian Industry and Agriculture*, particularly Chs. VI–VIII.

[22] Granick, *op. cit.*, Chs. IX, X.

[23] Also, the enterprise manager is hampered in economizing wages by the fact that the economic plan confron. him with not one target for the wage bill and the labor force but separate targets for several major categories of labor. But presumably this restraint is of very limited import. Among other things, under the complex Soviet planning system, the plant manager has a good deal of say about the formulation of the plan itself.

[24] See Bergson, *The Structure of Soviet Wages*, pp. 33 ff.

open market of limited dimensions in the form of the retail collective farm market and also the so-called state commercial shops, where goods were sold in unlimited amounts at high prices.

(iv) According to established policy and procedures, wages in the USSR are differentiated on much the same basis as under capitalism. Generally, as under capitalism, higher wages are paid for more arduous, more complex, and more responsible jobs. Piece work is applied extensively.[25]

According to an investigation by the writer, wage differentials prevailing in the USSR under the five year plans have been comparable in magnitude to those prevailing in Tsarist Russia and possibly also in the United States.[26]

### COLLECTIVE FARM INCOMES

As in the case of wage differentials, so in the case of the relation of farm incomes and industrial wages, this study assumes provisionally no major divergence between ruble prices and Adjusted Factor Cost.

Seemingly the requirements for a correspondence of ruble prices and Adjusted Factor Cost are reduced here essentially to one: "real" incomes on the farm must be the same as those in industry for workers of comparable skill; or at least this must be so except for differences in Disutility. Referring, then, to this requirement, the evidence for the assumption just stated consists chiefly of some data the writer has compiled concerning the comparative real incomes of collective farmers and industrial workers. The data are set forth below. Despite their obvious limitations, the conclusion is indicated that in regard to real income the collective farmer may have compared favorably in 1937 with the industrial worker. It must be recalled here that 1937 was a particularly prosperous year in Soviet agriculture, but the comparison is still illuminating on the question under consideration; the more so in view of the common assumption that the Soviet government systematically discriminated against the farmer and in favor of the industrial worker.

(i) According to calculations set forth in Appendix G, the collec-

[25] *Ibid.*, particularly Chs. XI and XII.   [26] *Ibid.*, Chs. V–VIII.

tive farmers of the USSR worked on and off the farm some 32 million man years in 1937, for which they earned about 56 billion rubles, or an average of 1,760 rubles per man year. The man year referred to is a theoretic full-time man year of 280 days; actually, able-bodied male collective farmers worked on the average about 270 days in 1937; and male and female collective farmers together, 235 days. The cited figure on collective farm employment represents the result of a collation of various Soviet collective farm employment data, which in turn are based on collective farm annual accounts and sample studies. The indications are that the figure is an understatement. The total collective farm earnings are calculated from data in Table 1, p. 18, together with other information from Soviet sources.

(ii) Collective farm earnings as just calculated include farm income in kind valued at farm prices, gross of subsidies, and net of turnover taxes. For present purposes, however, farm income in kind evidently should be valued at retail prices, or strictly speaking at retail prices less an allowance for home processing costs. As to how large a spread there is in the USSR between farm and retail prices, the figures in the turnover taxes cited above, p. 60, taken together with the estimate of subsidies on farm produce, provide some indication. If account is taken only of the taxes levied at the procurement stage, and farm income in kind is valued at farm prices before subsidies and gross of turnover taxes, the average income of the collective farmer is increased from 1,760 to about 2,240 rubles per man year.[27] At retail prices, the corresponding figures would still be considerably higher.

(iii) According to data compiled in current Soviet statistical reports, the average earnings of workers employed in industry amounted to 3,005 rubles in 1937. This refers to all types of employees:

[27] The total marketed share of farm produce in 1937 amounted to 38.0 billion rubles. This includes the government procurements at procurement prices amounting to 15.7 billion rubles; collective farm sales amounting to 17.8 billion rubles; and subsidies amounting to 4.5 billion rubles. (See Appendix A, p. 105.) The turnover taxes paid by procurement agencies amounted to 24.1 billions. Hence the inclusion of taxes and deduction of subsidies would raise the average price of farm produce by 52 percent, i.e., $(24.1 - 4.5) \div 38.0$. Farm income in kind, initially 29.4 billion rubles, is now increased to 44.7 billion rubles, and total collective farm earnings are raised correspondingly.

ordinary workers of a rank below foreman; apprentices and service personnel; and engineers, technical, managerial, and other white collar workers. The corresponding figure for ordinary workers of a rank below foreman was 2,820 rubles.[28] On the average, the Soviet industrial worker was employed 266 days in 1935.[29] As has been mentioned, the Soviet current statistical reports on wages are not entirely comprehensive with respect to either the labor force or wage payments. For the workers covered, however, it is doubtful that the understatement in average wages on this account is more than a few percent.[30]

(iv) Around the time considered, the general level of prices in the state and cooperative retail shops probably was somewhat higher in the country than in the city. In the case of most basic foods (bread, sugar, meat, lard, butter, fish, etc.), prices in the country were the same as in the city. In the case of a number of manufactured goods (cotton, woolen, and linen textiles, clothing, etc.), however, the rural prices were several percent higher than those in the cities, reflecting a higher trading margin allowed in the rural shops.[31]

[28] Gosplan, *Tretii piatiletnii plan,* pp, 228–29. Industry, as the term is used in Soviet statistics, includes manufacturing, mining, and electric power generation.

[29] TSUNKHU, *Trud v SSSR* (Labor in the USSR), p. 78.

[30] Mention should be made here also of the income in kind which Soviet employees obtained from domestic gardens. According to our computations, based on Soviet data, the aggregate net output of these gardens for all Soviet employees constituted something less than 10 percent of the net farm produce at the disposal of all households together in 1937. Presumably this income in kind accrued mainly to employees in rural localities, who were chiefly in non-industrial pursuits, and particularly to those engaged in the state farms, machine-tractor stations, and in lumbering, for whom it undoubtedly was a very sizable supplement to their rather low money wages: 2121 rubles a year for those in agriculture and 1920 rubles a year for those in lumbering.

[31] Thus, in the case of cotton cloth, the trading margin in the urban localities varied, depending on the region, from 2.9 to 6.2 percent of the wholesale price and in rural localities from 7.0 to 12.4 percent of the wholesale price. In the case of many products, including some foods such as salt, prices were differentiated also to allow the rural shops located at a distance from railway depots to cover the cost of long-distance truck hauling. See G. I. Kuznetsov, editor, *Sbornik otpusknykh i roznichnykh tsen i torgovykh nakidok na prodovol'stvennye tovary* (Handbook of Wholesale and Retail Prices and Trade Margins on Foodstuffs); *idem, Otpusknye i roznichnye tseny i torgovye nakidki na promtovary* (Wholesale and Retail Prices and Trade Margins on Industrial Goods). While the collective farmers thus were at a slight disadvantage, in comparison with city workers, in their purchases in the state and cooperative retail shops, it should be noted that they profited from the practice of collective farms of selling produce, particularly animal products, to their members at special low prices. In their purchases of animal products from the collective farms in 1935, collective farm

(v) No information is at hand on the comparative availability of goods in the urban and rural retail shops.

(vi) Owner-occupancy of housing is quite general in the country but of comparatively minor importance in the city. If the services of such housing are valued and added to earned income, then the increment is on the average greater for the collective farmers than for the industrial workers. As has been indicated (Table 1), if the evaluation is at the very low rentals established for the government housing, the sums involved are necessarily quite small, so that our comparison is little affected. In a more accurate calculation, however, both the urban and rural housing, rented and owner-occupied, evidently would have to be revalued, and as a result any differences in the average amount of space available might emerge as a significant cause of differentiation in real incomes. Considering the cumulative deficiencies in urban housing construction under the five year plans, it is entirely possible that as of 1937 this factor would have operated favorably to the country.[32]

In regard to the relation of farm income to wages, as in regard to wage differentials, reference should be made also to the organization of the labor market. In the interwar period there was freedom of choice of occupation not only for workers in industry; farmers too had such freedom and, in particular, had open to them the opportunity to move to the city. Under the five year plans new recruits for industry apparently were obtained to a considerable extent as a result of the voluntary decisions of farm workers so to move.[33] At the same time, the unemployment that prevailed in the cities in the late twenties was liquidated in the First Five Year Plan, so there is fur-

members realized an economy of 110 million rubles in comparison with what they would have paid in the retail collective farm market. See M. Nesmii, "Dokhody kolkhozov i kolkhoznikov" (Incomes of Collective Farms and Farmers), *Planovoe khoziaistvo* (Planned Economy), 1938, No. 9 p. 87.

[32] For present purposes, interest seems to center on comparative *real* earnings, exclusive of free social services, rather than on comparative *real* incomes, inclusive of such services. It should be noted, however, that in regard to such services, particularly health care and education, the urban worker undoubtedly was much better off than the farmer.

[33] Again, as in the case of the industrial workers, there were controls, but it appears that until 1940 these, too, were of a relatively limited sort. See Bergson, *op. cit.*, pp. 143 ff.; and G. Bienstock *et al.*, *Management in Russian Industry and Agriculture*, pp. 38–39.

ther reason to believe that rural-urban income differentials were more or less on a par. Of course, the possibility must be reckoned with that there was "disguised" unemployment on the farms, in the sense that at the prevailing differentials many farmers would have preferred to work in the city but were deterred from doing so by the lack of openings there. One wonders, however, whether such disguised unemployment could assume any large proportions without there being first some sizable amount of unemployment in undisguised form. Actually there is some evidence that, to the contrary, the labor recruitment quotas of industry were not being entirely filled towards the end of the Second Five Year Plan; [34] and the possibility suggests itself that the prevailing income differential may even have been inadequate to maintain the flow of labor from country to the city. In this connection, it should be observed that the parity in rural and urban real incomes in 1937 suggested by our calculations in all probability would have represented a relative gain for the peasant in comparison with the situation on the eve of the five year plans.

It already has been mentioned that the government introduced extensive labor controls in 1940. These included a draft of urban and farm youth for vocational training and subsequent administrative allocation to industry. Although international conditions no doubt played a part, one is impelled to think that adverse longer-term market forces of the sort just described may also have influenced the government's decision in this regard. Interestingly, while other post-1939 measures like rationing have by now been abandoned, the labor controls of 1940 still are operative.[35]

[34] M. Sonim and B. Miroshnichenko, *Podbor i obuchenie rabochikh kadrov v promyshlennosti* (Selection and Training of Labor Cadres in Industry), p. 4; S. Trubnikov, "Istochniki komplektovaniia rabochei sily v SSSR" (Sources of Recruitment of the Labor Force of the USSR), *Problemy ekonomiki* (Problems of Economics), 1936, No. 6, p. 150.

[35] Of course, regardless of the causes of the labor shortages, the possibility was open to the government to increase the urban-rural income differential in order to stimulate the labor flow to industry once again. But an increase in urban living standards to this end necessarily would have meant a reduction in investment and in the tempo of industrialization. On the other hand, these consequences would have been avoided by reduction in rural living standards, but the government might have considered such a step as politically less appealing than the labor draft.

I have left open here the question as to what extent any equalization of urban and

For a correspondence of ruble prices and Adjusted Factor Cost, the requirement evidently is that farm incomes *qua* returns to labor as distinct from land or capital should correspond to the earnings of comparable industrial labor. As has been mentioned, however, collective farm incomes in the USSR include some element of differential rent. No doubt they also include a return on the collective farmers' private capital. The import of these complexities for our ruble in-

---

rural real incomes that occurred under the five year plans represented a decline in urban and to what extent if at all an increase in rural real incomes. My impression is that it was in part if not altogether the former (see p. 10, note 11). But very possibly the government considered *this* politically expedient in view of the fact that many of the urban workers were coming from the country where to begin with living standards were even lower.

Readers familiar with Soviet economic history will not fail to note the interesting analogy between the labor problem of the late thirties as just defined and the grain problem that prevailed on the eve of the First Five Year Plan. In both cases the key to the problem seems to be the policy of high tempo industrialization, leading in one instance to a shortage of labor and in the other to a shortage of grain. In both cases the government at least in theory faced a choice between a policy of relying on the price system and voluntary recruitment and one of relying on more direct administrative controls. Both collectivization and labor conscription meant a decision in favor of the latter.

Of course, to speak at all of a shortage of industrial labor has a rather paradoxical character in Soviet conditions. On the eve of the five year plans, Soviet economists considered the problem to be more one of an *excess* in rural labor rather than one of a *shortage* in urban labor; and while there was in fact a large migration to the cities thereafter, the Russian population was still predominantly rural before the war. But the suggested trends in income differentials might go far to resolve the paradox.

Finally, the view set forth here on the prewar labor market and the introduction of controls finds some support in Soviet analyses; for example (though subject to an obvious reservation), the following statement made by Stalin in March, 1939: "Now it is no longer a question of finding room in industry for unemployed and homeless peasants who have been set adrift from their villages and live in fear of starvation—of giving them jobs out of charity. The time has long gone by when there were such peasants in our country. And this is a good thing, of course, for it testifies to the prosperity of our countryside. If anything, it is now a question of asking the collective farms to comply with our request and to release, say, one and a half million young collective farmers annually for the needs of our expanding industry. . . . The collective farms are quite able to meet this request of ours, since the abundance of machinery in the collective farms releases a portion of the rural workers, who, if transferred to industry, could be of immense service to our whole national economy." (J. Stalin, *Leninism: Selected Writings,* p. 455. See also Nancy Baster, *Agrarian Overpopulation in the USSR, 1921–1940,* unpublished essay on file at the Russian Institute, Columbia University, 1949, pp. 76–77.)

On the other hand, my interpretation is in contrast to one frequently expressed by non-Soviet economists, namely, that the labor legislation of 1940 was purely an emergency program. See M. Dodd, *Soviet Economic Development since 1917,* pp. 447–48.

come figures, however, is perhaps properly left to the reader's judgment.

## NATIONAL INCOME AT ADJUSTED FACTOR COST

I have said I would appraise here and try to correct for the effects of divergencies between ruble prices and Adjusted Factor Cost on two tabulations, those of the national income by use and by industrial origin. The former tabulation is shown in Table 8, columns 1 and 2 and the latter in Table 9, columns 1 and 2.

To refer first to the divergence due to the turnover tax, fortunately for our purposes it is possible to appraise quantitatively and hence discount for the effect of this factor. Thus in Table 8, columns 3 and 4, data are presented on the incidence of the tax by use category, and in Table 9, columns 3 and 4, on the incidence of the tax by economic sector. In the case of the breakdown in terms of use categories, the figures are intended to show the total incidence of the tax, both direct and indirect; e.g., under consumption are included not only the taxes on jam but also on sugar.[36] Evidently these calculations provide a basis to adjust the figures on the national product so as to allow for the distortion in the relation of prevailing ruble prices to Adjusted Factor Cost due to turnover tax.

The computation of the incidence of the turnover tax by use categories and economic sector is described in Appendix E, but it may be advisable to say a word about this here. The breakdown by industrial sector is readily made, and I believe with a minimum of error, from Soviet data on the breakdown of the turnover tax by economic commissariat. The computation of the incidence of the tax by use category is by contrast subject to an appreciable margin of error. It starts with comparatively detailed estimates of the breakdown of the national product by use. Thus, estimates are made of the breakdown of government expenditures as between expenditures on commodities and expenditures on services; the breakdown of investments as between working and fixed capital, and so on. It is known that cer-

---

[36] The Russians make a point of the fact that they avoid multiple taxation insofar as they levy the turnover tax only once on a given product. While this seems in fact to be the case for any given product, there is much duplication as a result of the conjuncture of taxes on raw materials and finished products.

## TABLE 8

### GROSS NATIONAL PRODUCT BY USE, IN TERMS OF ESTABLISHED AND ADJUSTED PRICES, USSR 1937 [a]

| ITEM | VALUE OF PRODUCT AT ESTABLISHED PRICES | | CALCULATED TURNOVER TAX | | SUBSIDIES | | VALUE LESS TURNOVER TAX PLUS SUBSIDIES | | ADJUSTMENT FOR FARM PRICES | VALUE AT ADJUSTED PRICES | |
|---|---|---|---|---|---|---|---|---|---|---|---|
| | Billion rubles | Percent | Billion rubles | Percent | Billion rubles | Percent | Billion rubles | Percent | Billion rubles | Billion rubles | Percent |
| | (1) | (2) | (3) | (4) | (5) | (6) | (7) | (8) | (9) | (10) | (11) |
| 1. Consumption of households | 183.5 | 62.9 | 61.1 | 80.5 | 4.2 | 52.5 | 126.6 | 56.5 | (—)1.9 | 124.7 | 55.7 |
| 2. Communal services | 27.4 | 9.4 | 4.8 | 6.3 | .5 | 6.3 | 23.1 | 10.3 | .7 | 23.8 | 10.6 |
| 3. Government administration, including NKVD | 7.4 | 2.5 | .5 | .7 | 0 | 0 | 6.9 | 3.1 | 0 | 6.9 | 3.1 |
| 4. Defense | 17.5 | 6.0 | 1.0 | 1.3 | .7 | 8.8 | 17.2 | 7.7 | .1 | 17.3 | 7.7 |
| 5. Gross investment | 56.1 | 19.2 | 8.5 | 11.2 | 2.6 | 32.5 | 50.2 | 22.4 | 1.0 | 51.2 | 22.9 |
| 6. Gross national product | 291.8 | 100.0 | 75.9 | 100.0 | 8.0 | 100.0 | 223.9 | 100.0 | 0 | 223.9 | 100.0 |

[a] Sources and methods are set forth in Appendix E. Minor discrepancies between indicated and calculated totals are due to rounding.

## TABLE 9

### NET NATIONAL PRODUCT BY ECONOMIC SECTOR, IN TERMS OF ESTABLISHED AND ADJUSTED PRICES, USSR 1937 [a]

| ITEM | NET NATIONAL PRODUCT | | TURNOVER TAXES | | SUBSIDIZED LOSSES | | NET NATIONAL PRODUCT LESS TURNOVER TAXES PLUS SUBSIDIES | |
|---|---|---|---|---|---|---|---|---|
| | *Billions of rubles* (1) | *Percent* (2) | *Billions of rubles* (3) | *Percent* (4) | *Billions of rubles* (5) | *Percent* (6) | *Billions of rubles* (7) | *Percent* (8) |
| 1. Agriculture | 60.7 | 21.2 | | | 4.5 | 56.3 | 65.2 | 29.9 |
| 2. Industry and construction | 121.2 | 42.4 | 47.7 | 62.8 | 2.5 | 31.3 | 76.1 | 34.9 |
| 3. Transportation and communications | 16.3 | 5.7 | | | .5 | 6.3 | 16.8 | 7.7 |
| 4. Trade, including restaurants | 37.5 | 13.1 | 27.6 | 36.4 | .5 | 6.3 | 10.4 | 4.8 |
| 5. Finance | 1.9 | .7 | | | | | 1.9 | .9 |
| 6. Services, including government | 38.0 | 13.3 | | | | | 38.0 | 17.4 |
| 7. Other | 1.1 | .4 | | | | | 1.1 | .5 |
| 8. Statistical discrepancy | 9.4 | 3.3 | .7 | .9 | | | 8.7 | 4.0 |
| 9. Total | 286.0 | 100.0 | 75.9 | 100.0 | 8.0 | 100.0 | 218.1 | 100.0 |

[a] For sources and methods, see Appendix C. Minor discrepancies between computed and indicated totals of items are due to rounding.

tain use categories, most importantly services, are entirely free of the tax. Referring only to categories subject to the tax, it was found that the average effective rate of the turnover tax in all such categories together was 35.1 percent (75.9 billion rubles of turnover taxes divided by 216.5 billion rubles of national product subject to the tax). For purposes of the calculation, the average rate itself was taken as applying to a number of use categories, including commodities used in the provision of communal services, government administration, NKVD, army subsistence, and working capital investment. Except for the taxes on retail sales, which were computed as a residual, the taxes on other categories were estimated on the basis of scattered information, including particularly the official turnover tax schedules by commodities. The results are in broad agreement with the breakdown of the turnover tax by economic commissariat referred to above.

A calculation has been attempted also of the incidence of subsidies. The results are shown in Table 8, columns 5 and 6, and in Table 9, columns 5 and 6.

Details of this calculation are in Appendix E. Because of Soviet reticence on subsidies, it involved a good deal of guesswork, and in the case of the allocation by economic sector I am in doubt even as to the sectors affected. Clearly, however, as is assumed here the bulk of the subsidies fall on industry and agriculture. In the case of industry, the subsidies represent for the most part the appropriations to cover losses still imposed on many basic industries after the price reform of April, 1936. In the case of agriculture, they represent chiefly appropriations to the MTS needed to compensate for the undervaluation of income in kind at government procurement prices. Probably prices were too low and hence subsidies were paid also for the state farms. The estimate of subsidies to transport is intended to take into account numerous Soviet reports of losses in water transport, and also the possibility of losses in other sectors. It is believed, however, that at the time studied, the railroads made their own way financially. The estimate of subsidies to trade allows for the appropriations, referred to earlier, which compensate procurement agencies for premiums paid farmers for extra-norm deliveries of certain farm products and for certain additional expenses.

In Table 8, columns 7 and 8, are set forth data on the national product by use after the deduction of the turnover tax and addition of subsidies. In Table 9, columns 7 and 8, are corresponding data for the national product by economic sector.

The data in column 9 of Table 8 bear only in part on the turnover tax and subsidies, but it is just as well to refer to them at this point. The deduction of the turnover tax from and addition of subsidies to the values of goods and services allocated to different uses evidently amounts to a revaluation of these goods and services. Seen from this viewpoint, however, the calculation is open to criticism in one particular. It happens that retail collective farm market sales are free of the turnover tax and subsidies. Under the heading of consumption, the turnover tax and subsidies fall almost exclusively on the goods sold at retail in state shops.[37] Accordingly, while the adjustment for taxes and subsidies means that the prices of the state shops goods are changed—on balance there is a reduction—comparable goods sold on the retail collective farm market continue to be valued at their old prices. Obviously logic requires that the retail collective farm market prices be reduced in the same proportion as those of the state shops.

The adjustment for farm prices in column 9 is intended in part to measure the effect of such a reduction in retail collective farm market prices and in part the effect of an *increase* in average government farm procurement prices now to be explained. Since elements of farm income, aside from taxes and subsidies, are a matter for separate consideration, the adjustment in column 9 has been made on the assumption that farm income is constant. The reduction in collective farm market prices means a reduction in farm income; hence for the purposes of column 9, I assume provisionally an increase in average government farm procurement prices such as to offset this decline in income. It should be observed that so far as the different use categories in Table 8 are concerned, the reduction in collective farm market

[37] Actually, there was at the time studied a tax of 3 percent on certain retail collective farm market sales, particularly sales in organized trading outlets such as shops and booths as distinct from sales from carts and the like. Just how extensive such sales were is not known; probably they were a minor proportion of the total. In any event, I disregard taxes on retail collective farm market sales throughout this study.

prices is reflected only in the value of consumption, while the increase in farm procurement prices shows up also in other categories. Necessarily the determination of the incidence of this increase was on a rule-of-thumb basis.

The data in Table 8, columns 10 and 11, represent the breakdown of the national product by use after all adjustments. This tabulation is referred to hereafter as being in terms of "adjusted rubles." So far as retail prices are concerned (in both state shops and collective farm markets), the adjustments involve a reduction of 45.4 percent.[38]

An adjustment of the sort made in Table 8, column 9, it should be evident is not required in Table 9. The changes in collective farm market and government procurement prices are entirely internal to the agricultural sector, and leave the values produced by all the different economic sectors, including agricultural, unchanged. Accordingly, the data in Table 9, columns 7 and 8, are in terms of the same adjusted ruble prices as those in Table 8, columns 10 and 11.

As a result of the Soviet treatment of the net charge for capital, there are two sources of a divergence between ruble prices and Adjusted Factor Cost to be considered: that due to the omission of such a charge and that due to the inclusion of a possibly arbitrary charge under the heading of profits.

It is believed that these divergencies are of relatively limited importance for our calculations. On the one hand, it is necessary to deduct

---

[38] I have said that the decline in collective farm market prices is assumed proportional to that in state retail shop prices. It may be in order to note that for purposes of the computation in terms of adjusted rubles the decline in state retail shop prices is itself taken to be the net resultant *after allowance for both the elimination of the turnover tax and subsidies and the rise in procurement values needed to offset the decline in collective farm market prices.* This is computed at once from the following equation:

$$k \text{ x 111.5 bil. rubles} = 55.9 \text{ bil. rubles} - .727 \text{ x } k \text{ x 16.0 bil. rubles}$$

Here $k$ is the decline in state retail prices; 111.5 billion rubles is the turnover in state retail shops; 55.9 billion rubles is the calculated amount of the turnover tax in excess of subsidies falling on state retail sales; .727 is the calculated proportion of the rise in procurement values which falls on state retail sales; and 16.0 billion rubles is the amount of collective farm market sales. This computation evidently avoids the inconsistency that would result if the decline in state shop prices were figured on the basis of the elimination of the turnover tax and subsidies alone and collective farm market prices were assumed to decline by the same amount, without allowing for the fact that the corresponding increase in procurement values ultimately would mean somewhat higher retail prices.

the part of net profits as now recorded that is over and above wages of management. Including the net income of collective farms not distributed among the membership, the total net profits recorded amounted to 21.6 billion rubles in 1937. Of this it is estimated that somewhat less than a quarter accrued to agriculture and over a half to industry.[39] On the other, it is necessary to allow for a return on capital. The indications are that the total stock of fixed capital in the USSR in 1937 might possibly have been of the order of 325 billion rubles in terms of the prices of that year.[40] At a rate of return of say 8 percent, this would mean a total yield on capital of 26 billion rubles. In the light of available data on the industrial disposition of Soviet capital,[41] I believe this total would be divided among the different economic sectors in proportions broadly similar to the profits now recorded there. The upshot of such adjustments very likely would be relatively significant in the case of some of the lesser sectors, particularly transportation, where the recorded profits (about 6 percent of the total for the whole economy) clearly are well below the average in relation to fixed capital. But in respect to agriculture and industry there hardly could be a change of more than one or two percentage points either way.[42]

For purposes of illustrating the possible effect on the tabulation by use, I assume (as may not be entirely wide of the mark) that, while the profit charge presently recorded in investment amounts to only one fifth of the 21.6 billion ruble total for all sectors,[43] the correct

[39] See Appendix C.

[40] According to N. Voznesenskii, *Voennaia ekonomika SSSR v period otechestvennoi voiny* (War Economy of the USSR in the Period of the Patriotic War), p. 12, the fixed capital of the USSR, exclusive of cattle, amounted to 564 billion rubles in 1937 in terms of 1945 prices. According to calculations of N. Jasny (*Soviet Prices of Producers' Goods*, p. 20), the costs of all capital investments increased about 70 percent from 1937 to 1945.

[41] See Central Administration of Economic and Social Statistics, USSR, *Socialist Construction in the USSR*, p. 12.

[42] Needless to say, this calculation based on an assumed yield of 8 percent is intended only as illustrative, but perhaps it is permissible to suppose that the assumed yield is of the right order of magnitude so far as concerns the productivity of capital in the sense which our Adjusted Factor Cost Standard implies. Also, one might think of the prevailing time preference in the same terms, though of course reference must be to the "planner's" rather than the "consumer's" preferences.

[43] See the data on profits in Appendix C, pp. 126 ff., and other information on the same subject in the sources cited there.

figure would be one third of the alternative total of 26 billion rubles considered above.[44] Taking as a point of departure the tabulation in adjusted rubles, this would mean an increase in the share of gross investment in the gross national product from 22.9 to 24.4 percent, and in the share of net investment in the net national product from 20.8 to 22.4 percent.

Insofar as depreciation is understated in our calculations, what is called for, for present purposes, is an upward revision of prices all around, in order to provide for an increased depreciation charge. The increase in the prices for any industry would depend on the amount of fixed capital employed. In the case of the tabulation by sector, where the figures show a value added *net* of depreciation, obviously such a price adjustment would have no effect.

In the case of the tabulation by use, presumably the main effect would be to increase gross investment and reduce net investment in relation to consumption. (While investment goods prices rise more than those of consumers' goods, the entire increase in depreciation for the whole economy must be deducted from the revised *gross* to obtain the revised *net* investment.) But again these changes would have to be limited. The aggregate of current depreciation charges for the USSR in 1937, as recorded in Table 2, amounted to 5.8 billion rubles. This is about 2 percent of the gross national product in prevailing rubles and 2.6 percent of the gross national product in adjusted rubles. According to the calculations of the U.S. Department of Commerce, the current depreciation charges for the United States in 1937 come to 6.8 billion dollars, or 7.6 percent of the gross national product. Let us suppose that the correct figure for the USSR would be double the present one, or 5 percent of the gross national product in adjusted rubles. (In relation to national income, the total stock of fixed capital in the USSR presumably is less than in the United States, so depreciation should be less also.[45]) Suppose also as

[44] See the data on fixed capital in Central Administration of Economic and Social Statistics, *op. cit.*, p. 12; P. A. Khromov, *Amortizatsiia v promyshlennosti SSSR* (Depreciation in Industry of the USSR), pp. 18–19.

[45] I have cited above a figure of 325 billion rubles for the total amount of Soviet fixed capital in 1937. This amounts to 1.1 times the gross national product in prevailing rubles and 1.45 times the gross national product in adjusted rubles. For the United States there is at hand an estimate of fixed capital for the year 1935, advanced by the·

was implied in the calculation made earlier on interest that one third of the fixed capital of the country was used to produce investment goods. Still taking the tabulation in adjusted rubles as the point of departure, the postulated increase in depreciation means that gross investment would increase from 22.9 to 23.1 percent of the gross national product and net investment would decline from 20.8 to 19.0 percent of the net national product.

It has been mentioned that the turnover tax is in some measure a charge for agricultural rent. Insofar as this is so, the deduction of the full amount of the turnover tax leads to an undervaluation of agricultural production from the standpoint of the Adjusted Factor Cost Standard. A question might be raised, too, about the extent to which the incomes actually accruing to collective farmers are in accord with this standard. The evidence is against any major divergence between prices and Adjusted Factor Cost on this account, but obviously on all counts the extent to which agricultural prices are in accord with Adjusted Factor Cost is rather conjectural.

Under the circumstances, some interest may attach to a further calculation that the writer has made. This is intended to illustrate the effects of our national income tabulations of a hypothetical increase in agricultural procurement prices by 25 percent, due to a revaluation of farm produce. It is supposed that collective farm market prices are in a constant relation to state retail prices, and that both together increase as a result of the increase in average procure-

---

National Resources Committee (*The Structure of the American Economy*, Part I, Washington, D.C., June, 1939, pp. 374–77): 190 billion dollars, or 2.6 times the gross national product of the United States in that year. For obvious reasons it would be more in order here to cite a figure for the United States for, say, the year 1937 when capital in the United States was more fully utilized than for the year 1935, when below capacity operations must have still been fairly extensive. On this basis, the ratio of capital to income for the United States presumably would be less than indicated.

The figure cited for Soviet fixed capital probably is comparable to that for the United States, insofar as it apparently includes housing as well as productive fixed capital. It might be supposed, too, that the Soviet figure, like that for the United States, is net of depreciation, but unhappily this important question is conjectural in Soviet sources.

It should be noted that the Soviet labor force in 1937 amounted to some 70 millions, while the United States labor force in 1935 amounted to 42 millions. Other things being equal, this disparity in itself should have resulted in a smaller capital-income ratio for the USSR than for the United States.

ment prices by 25 percent. Necessarily the main impact of the revaluation falls on consumption, but there are also limited effects in other use categories where agricultural produce is involved. Starting with the tabulation of the national product by use in terms of adjusted rubles (Table 8, columns 10 and 11), the upshot of the revaluation is that consumption increases from 55.7 to 57.5 percent of the gross national product and gross investment declines from 22.9 to 22.0 percent. Net investment decreases from 20.8 to 20.1 percent of the net national product. The reader can surmise the general effect of the revaluation on the breakdown of the national product by economic sector.[46]

[46] Still another calculation may be of interest in connection with this question of the valuation of agricultural income:

(i) According to Soviet official data, the gross harvest of agriculture in 1937 amounted to 20.1 billion rubles in terms of 1926–27 prices (TSUNKHU, "Sotsialisticheskoe sel'skoe khoziaistvo Soiuza SSR," as reprinted in *Planovoe khoziaistvo,* 1939, No. 7, p. 160.) Like all Soviet production figures in 1926–27 prices, this one on agriculture presumably is subject to a margin of error because of deficiencies in Soviet statistical methodology already mentioned. (See above, p. 5.) It is doubtful, however, that the deficiency and hence the error was nearly as serious for agricultural production as for, say, that of industry.

(ii) Allowing for production expenses as independently estimated (see Appendix A, pp. 105 ff.), this means that the net national product of agriculture was 12.7 billion rubles in terms of 1926–27 prices.

(iii) According to Soviet data, retail food prices in 1928 averaged 112 percent of 1926–27: Tsentral'noe Statisticheskoe Upravlenie (Central Statistical Administration), *Kontrol'nye tsifry 1928–1929 gg.* (Control Figures, 1928–1929), Moscow, 1929, p. 501; and *idem, Statisticheskii spravochnik SSSR za 1928* (Statistical Handbook of the USSR for 1928), Moscow, 1929, p. 725. According to calculations in S. N. Prokopovicz, *Quarterly Bulletin of Soviet Russian Economics,* May, 1941, p. 130, retail food prices were at the following levels in specified years (1913 = 100):

| 1928 | 182.0 |
|------|-------|
| 1936 | 1553.3 |
| 1938 | 1600.8 |

Taking the corresponding index for 1937 as an average of the indices for 1936 and 1938, the retail food price level for 1937 is 866 percent of that for 1928. Because of the limited number of commodities taken into account—eleven in all—and the difficulties that must have been encountered in obtaining comparable price data for different years, Prokopovicz' calculations must be subject to a sizable margin of error. These figures indicate that as of 1937 retail food prices were 970 percent of 1926–27. As was indicated above, the adjustment for the turnover tax in 1937 results in a reduction in retail prices generally by 45.4 percent. Assuming this applies to food prices, the conclusion is that *adjusted* retail food prices as of 1937 were 529 percent of actual retail food prices in 1926–27.

While farm procurement prices vary, through the use of turnover taxes and also some subsidies, prices to processors are made more or less uniform. In adjusting for the turnover tax and subsidies, then, are we not left again with multiple prices and, consequently, a new distortion in the relation of prices and Adjusted Factor Cost? The answer briefly is this: So far as concerns the tabulation by economic sector, the adjustment obviously must leave the balance of value added as it was established originally on the basis of uniform material prices. So far as concerns the tabulation by use, I hardly need say that in adjusting for the turnover tax and subsidies no account could be taken of the variation in the tax in dependence on the type of the procurement or of the restriction of subsidies to higher priced procurements: accordingly, the deduction of the tax should leave the prices to processors no less uniform than they were to begin with.[47]

While the prices of agricultural products generally are made uni-

---

(iv) If farm prices had increased in this same proportion, the net national product of agriculture would have amounted to 67.1 billion rubles in 1937 (i.e., 12.7 billion rubles x 529 percent). According to data compiled previously (Table 9) the net national product of agriculture in terms of the adjusted farm prices of 1937 amounted to 65.2 billion rubles.

While as has been noted this foregoing calculation is subject to error at every stage, it is somewhat reassuring in regard to the valuation that has been placed on farm income. What is implied, of course, is that once the turnover tax and subsidies are eliminated from the price structure, farm prices on the average bore about the same relation to retail prices in 1937 as they did in 1926–27. In the earlier year, agriculture was still almost entirely in private hands; the government relied primarily on market processes rather than compulsion to meet its food requirements; and while there were limited indirect taxes the turnover tax was not yet in existence.

Insofar as average realized farm prices correspond to Adjusted Factor Cost, it follows—to revert to the question raised above, pp. 26 ff.—that it is correct to use these prices as has been done in this study in valuing farm income in kind. The money value of farm services—in terms of these prices—is low in relation to money wages in the city, and cannot serve to indicate the relation of farm to city real income, but since in the case of income in kind no transportation and distribution services are involved, this is entirely in accord with Adjusted Factor Cost. To the extent of the transportation and distribution services not rendered, the prices at which income in kind is valued must be below urban retail prices. The disparity is the greater if as is usually done the convention is adopted that home processing activities should not be included in national income.

[47] Once it is conceded that profits are conventional, it is difficult in any case to consider multiple prices as raising a special problem in regard to the determination of value added by sectors. But obviously without affecting the relation of profits to capital in any sector, the multiple prices might distort the relation of money values to Adjusted Factor Cost in the tabulation by use.

form by the use of taxes and subsidies, grain and possibly some other produce are nevertheless undervalued in stockpiles and exports, and accordingly in investment. The government appropriation for stock-piling of all sorts in 1937 amounted to 1.7 billion rubles,[48] or about three fourths of one percent of the gross national product in adjusted rubles. While the actual outlays undoubtedly exceeded this as a result of the bumper grain crop, presumably the undervaluation is of a magnitude no larger than others to which reference has been made.

It may be of value to tabulate here the various calculations that have been made regarding the share of investment in income:

|  | | GROSS INVESTMENT, PERCENT OF GROSS NATIONAL PRODUCT | NET INVESTMENT, PERCENT OF NET NATIONAL PRODUCT |
|---|---|---|---|
| (i) | Prevailing rubles | 19.2 | 17.6 |
| (ii) | Adjusted rubles | 22.9 | 20.8 |
| (iii) | Adjustment for 8 percent return on capital | +1.5 | +1.6 |
| (iv) | Adjustment for increase in depreciation charge from 5.8 to 11.6 billion rubles | +.2 | −1.8 |
| (v) | Adjustment for 25 percent increase in farm procurement prices | −.9 | −.7 |
| (vi) | Adjustment for undervaluation of stockpiles | +? | +? |

In view of the relatively nebulous character of the assumptions involved in calculations (iii) through (vi) the writer ventures to think the calculation in adjusted rubles should be accorded priority here, as a provisional representation of Adjusted Factor Cost. At the same time, the indications are that this latter calculation may err in some measure in the direction of understating investment and accordingly overstating consumption; at least this is so in the case of gross as distinct from net investment. In this connection, one should have in mind in addition to calculations (iii) through (vi) several deficiencies referred to previously: the inclusion in consumption rather than investment of farm household investment in kind, and the omission from investment of private investment in housing construction. In the adjusted ruble tabulation by use, presumably defense is understated along with investment. The adjusted ruble calculation by economic sector, it will be recalled, tends to understate some of the minor sectors, particularly transport.

[48] See Appendix B, p. 121.

RUBLE VERSUS DOLLAR VALUATION

In Table 10 are shown the following tabulations of the Soviet net national product by use:

(i) The tabulation for 1937 in terms of prevailing ruble prices that has been compiled in this study. This is taken from Table 8, columns 1 and 2, except that reference is now to the net instead of gross investment and national product.

(ii) The tabulation for 1937 in terms of adjusted ruble prices that has been compiled in this study. This is taken from Table 8, columns 10 and 11, except that reference is now to the net instead of gross investment and national product.

(iii) A tabulation of Colin Clark's for 1938 in terms of "international units." In this tabulation an international unit "is defined as the quantity of goods and services exchangeable for one dollar over the average of the years 1925–34." As is indicated, reference is to the volume of goods exchangeable for a dollar at American dollar prices; so the tabulation in effect involves the valuation of Soviet goods and services at average American dollar prices for the period 1925–34.

(iv) A tabulation of Julius Wyler's for 1937 in terms of United States dollar prices of 1940.

The apparent differences in the results of the different calculations as expressed in percentage terms no doubt reflect to an important degree purely statistical factors. The statistical limitations of our own tabulations already have been mentioned. With regard to the tabulations of Clark and Wyler, it should be noted that there is positive evidence that one or the other or both are subject to a large margin of error.[49]

There are reasons to think, however, that the differences in results

[49] On the basis of the U.S. Bureau of Labor Statistics index of wholesale prices, the figures on national income computed by Clark and Wyler have both been translated into United States dollars of the year 1937. The result for Clark is a figure of 30.6 billion dollars and for Wyler, a figure of 41.5 billion dollars. All the evidence points to an *increase* in Soviet national income from 1937 to 1938; so the discrepancy cannot be explained by the fact that Clark's calculation refers to the later year. Neither Clark nor Wyler, by the way, has as yet set forth the details of his calculations. In the case of Clark, however, details have been published on an earlier set of calculations in terms of British pound sterling prices. See his *Critique of Russian Statistics*.

## TABLE 10

## NET NATIONAL PRODUCT BY USE, USSR 1937, 1938

| | BERGSON, 1937 AT PREVAILING RUBLE PRICES OF 1937 | | BERGSON, 1937 AT ADJUSTED RUBLE PRICES OF 1937 | | CLARK,[a] 1938 IN INTERNATIONAL UNITS | | WYLER,[b] 1937 IN U.S. PRICES OF 1940 | |
|---|---|---|---|---|---|---|---|---|
| | Billion rubles (1) | Percent (2) | Billion rubles (3) | Percent (4) | Billion units (5) | Percent (6) | Billion dollars (7) | Percent (8) |
| 1. Consumption of households | 183.5 | 64.2 | 124.7 | 57.2 | 20.26 | 66.7 | 22.0[c] | 58.2[c] |
| 2. Communal services | 27.4 | 9.6 | 23.8 | 10.9 | } 6.38 | } 21.0 | } 6.9 | } 18.2 |
| 3. Government administration | 7.4 | 2.6 | 6.9 | 3.2 | | | | |
| 4. Defense | 17.5 | 6.1 | 17.3 | 7.9 | | | 4.3 | 11.4 |
| 5. Net investment | 50.3 | 17.6 | 45.4 | 20.8 | 3.74 | 12.3 | 4.6 | 12.2 |
| 6. Net national product | 286.0 | 100.0 | 218.1 | 100.0 | 30.38 | 100.0 | 37.8 | 100.0 |

a "Russian Income and Production Statistics," *Review of Economic Statistics*, November, 1947, p. 216.

b "The National Income of Soviet Russia," *Social Research*, December, 1946, p. 512.

c Includes social welfare expenditures of enterprises and trade unions, which in our calculations are treated as a part of communal services. It is not clear how this item is dealt with by Clark.

also reflect to some extent the difference in valuation procedures used. This of course is to be expected insofar as the ruble and dollar price structures differ. Thus it will be noted that the share of the national product invested is found by both Clark and Wyler to be less than in the adjusted ruble calculation. At the same time, Clark's (though not Wyler's) figure on the share of consumption is higher than ours. Bearing in mind the known facts about United States technical superiority over Russia, this pattern might readily be accounted for if it can be assumed that the United States technical superiority is more pronounced in the case of the more highly fabricated investment goods than in the case of the less highly fabricated consumers' goods, and that the adjusted ruble and dollar price structures reflect this difference.[50]

[50] Further light is shed on the possible extent of the difference in price structures by some striking calculations in Professor Gerschenkron's *A Dollar Index of Soviet Machinery Output, 1927-28 to 1937*, pp. 46 ff. For purposes of appraising the possible significance of his use of dollar prices as weights in an index of Soviet machinery output, Professor Gerschenkron makes a number of calculations concerning the effect of changes in weights on index numbers of American machinery output. Thus, he calculates the change in American machinery output from 1899 to 1939, using first dollar prices of 1899 and then dollar prices of 1939. Similar calculations are made with reference to the periods 1899-1923, 1909-1939 and 1909-1923. In all cases, the discrepancy due to the change in weights is startling. Thus, to cite the most extreme example, American machinery output in 1939 is 1,542 percent of 1899 if valuation is in terms of 1899 prices and 198 percent of 1899 if valuation is in terms of 1939 prices. The difference in results, of course, reflects a combination of differences in price structures with differences in the physical quantity regimen, but insofar as it reflects the former in part it is obviously suggestive also as to the possible magnitude of the difference between dollar and ruble price structures.

# 5 ECONOMIC IMPLICATIONS

INTRODUCTION

I come finally to the economic implications of our calculations. What is in mind falls under three headings to be considered in turn: prices and finance; resource allocation; and industrial structure. Where it seems of interest, comparisons are made with the United States. For the United States, reference is made throughout to the national income calculations of the U.S. Department of Commerce as recently revised.[1]

PRICES AND FINANCE

Under this heading, reference is more concretely to the households' propensity to save; the sources of Soviet finance; surplus value; and the relation of the government budget to the national income.

(i) *The Households' Propensity to Save.* As computed in Table 1, p. 18, the total savings of households come to 5.4 billion rubles, or 2.8 percent of their total income. For the United States, the corresponding figure is 5.3 percent for the year 1937, and 8.3 percent for the year 1946.[2]

Insofar as these data indicate that the household propensity to save in the USSR is less than in the United States, the result is readily understandable in terms of: (i) the much lower level of real income in the Soviet Union; (ii) the more comprehensive social insurance system in the USSR and the continuing full-employment which tend to obviate to some extent the need for personal savings;[3] (iii) the greater equality in the distribution of income in the USSR—a study by this writer[4] suggests that among wage earners alone, the inequality

---

[1] U.S. Department of Commerce, "National Income," Supplement to *Survey of Current Business,* July, 1947.

[2] *Ibid.,* p. 19. Taking income net of direct taxes, the proportions are: for the USSR in 1937, about the same as before, or 2.9 percent; for the United States in 1937, 5.5 percent, and in 1946, 9.3 percent.

[3] A question has been raised, however, by Manya Gordon (*Workers Before and After Lenin,* Section VII) as to the adequacy of the Soviet pensions and allowances.

[4] Abram Bergson, *The Structure of Soviet Wages.*

may be as great in the USSR as in the United States; there is hardly any doubt, however, that if all income recipients are considered, including recipients of unearned income in the United States, the inequality in the USSR would be less; and possibly (iv) the continuing inflation in the USSR under the five year plans, which constantly depreciates the value of accumulated savings (it is open to question, however, how sensitive the population is on this matter).

In view of these circumstances, however, a question might well be raised as to why the Russians save as much as they do, or indeed why they bother to save at all. As has been indicated, the actual savings are undoubtedly greater than the amount calculated here. No doubt there are definite reasons for the Russians to engage in some limited amount of savings; but note should be made here of the fact that the savings are not entirely voluntary. Apparently a fair proportion of the bond purchases reflect some degree of coercion.[5] Some of the

[5] In his *The Soviet Price System*, p. 153, Dr. N. Jasny makes the following comment concerning the present study in its earlier version: "One wonders what connection exists between Bergson's findings with reference to changes in incomes from 1928 to 1937 and his manner of treating Soviet institutions, which might have been in order in wartime, but seems out of place in 1950. Two examples will suffice. In 'Soviet National Income and Product in 1937. Part I' (p. 211), Bergson writes: 'The Soviet collective farm usually is described as a cooperative organization. The characterization is substantially but not entirely correct.' To whom does this word 'usually' in this statement pertain? Not to V. P. Timoshenko, Leonard Hubbard, William Lissner, Lazar Volin, and every other careful student of Soviet economy abroad. The collective farm of today is a state farm without a wage bill, a farm which forces the peasants to work and does not guarantee them a reward. The only co-operative feature of the collective farm is its name.

"In connection with Soviet loans, Bergson writes (*ibid.*, Part I, p. 234): 'Apparently a fair proportion of the bond purchases reflect some degree of coercion.' It is firmly established how many weeks' pay one has to sign up for. The purchasers of Soviet bonds have no right to sell them without permission, which is granted only in exceptional circumstances. The value of the bonds declines persistently because of inflation. In 1947 two-thirds of them were confiscated. . . . According to *Pravda* (Dec. 17, 1949), a man was accused illegally of buying 100 ruble bonds at 6 rubles and selling them at 10 rubles."

My statement on living standards in the earlier version of this study to which Dr. Jasny refers still appears in the present version on p. 10. I have already commented there on Dr. Jasny's criticism of this statement.

The reader will see from p. 12 above that I have reworded here the statement on the collective farm which Dr. Jasny quotes. This was to make it completely clear that I am using the term "cooperative" only in a formal legal sense. It is difficult to see, however, how anyone who reads in context the statement Dr. Jasny quotes could have been

savings, furthermore, may have been induced in some measure by shortages of one or another sort of consumers' goods at the established retail prices. For reasons stated in the previous chapter, it is believed that the open market functioned with reasonable efficiency in 1937, in the sense that there were not any general and serious deficiencies in available supplies at established prices. This does not preclude, however, that there were significant shortages in special lines, which to a limited extent would have resulted in forced savings.

(ii) *Sources of Soviet Finance.* Table 2, p. 20, shows at a glance all the sources of revenue utilized, as of 1937, to finance all the various activities the costs of which were not covered by sales to consumers: social and cultural projects, government administration, internal security and defense, and capital investments. The mainstay, evidently, was the turnover tax. This tax accounted for 65 percent of the aggregate revenue from all sources, and was alone more than sufficient to cover all requirements for capital investment.

As generally has been supposed, transfer receipts from the households, including direct taxes as well as savings, were strictly a minor element in the picture, accounting for only 8.0 percent of the aggregate revenue. According to the data in the table, the population was willing to put aside for the future less than one tenth as much as

---

misled on this aspect. The comments on the distribution of income and government controls and exactions, made above pp. 12 ff. and pp. 59 ff., all appeared in the earlier version. At the same time, it may be permissible to suggest that if Dr. Jasny himself had been seriously interested in ascertaining my views on the status of the Soviet collective farms he might have referred also to other writings of mine bearing more directly on this subject, particularly a review of his own *The Socialized Agriculture of the USSR* (*New Republic,* November 28, 1949). Thus, from this review he could have extracted the following quotation: "In sum, then, it would seem that up to the war collectivization was less the vehicle of economic progress that the Communists glowingly prophesied than an instrument of control, to bend peasant agriculture to the needs of the workers' state."

I am prepared to believe that a more intensive inquiry than I have felt able to make might possibly justify a stronger statement on the degree of coercion involved in government bond subscriptions than the one repeated in the text from the earlier version of this study to which Dr. Jasny objects. In this connection, however, it should be noted that the concern here is with the period around 1937. As Dr. Jasny omits to make altogether explicit, December, 1947, to which he refers, is the date when the government initiated a major monetary reform, involving among other things an exchange of bonds reducing the outstanding value by two thirds. The reform was the first of its sort in the USSR since the NEP.

the government was investing. This, of course, is a fact of first-rate political as well as economic significance.[6]

The large expenditures on social and cultural measures, defense, and capital investments necessarily involved the creation of consumers' purchasing power without any concomitant creation of consumers' goods. While showing the source of Soviet finance, Table 2 evidently sheds light at the same time on the government measures for dealing with this excess purchasing power. Again, the main factor was the turnover tax, which resulted in higher prices for the limited amount of consumers' goods available. According to a calculation in Appendix E, out of the total revenue of 76 billion rubles realized from the tax, something like 60 billion rubles were extracted directly or indirectly from the consumers' goods sold at retail by the state and cooperative shops. This amounted to 54 percent of the total turnover of these shops.

(iii) *Surplus Value*. As Soviet economists acknowledge, "surplus value" in a Marxian sense continues to exist in a socialist society, though accruing to the state rather than, as under capitalism, to private enterprise. In Table 3, p. 21, the item "Consolidated charges of government, etc." seems to provide a rough measure of the aggregate amount of surplus value in the USSR in the year considered: 102 billion rubles, or 36 percent of the *net* national product. To obtain a roughly comparable figure for the United States, the following items in the U.S. Department of Commerce calculations have been aggregated: rental incomes of persons, corporate profits, interest, and business taxes. For 1937 this sum amounts to $22.9 billions, or 28 percent of the net national product of this country.[7] With the inclusion of the incomes of unincorporated enterprises other than farms, the total comes to $29.6 billions, or 36 percent of the *net* national product.

[6] While it is conceded that in any case savings would be low in relation to investment, it has been suggested that the low level of savings in the USSR reflects in part the nature of the government's fiscal policies, and in particular that the households would save more if the government placed more reliance on direct rather than indirect taxes. Our impression is that this view is incorrect. While it is true that consumers' goods prices would be lower, consumption in *real* terms and hence voluntary savings in *real* terms necessarily would be about the same as before. As a result of the decline in prices, voluntary savings in monetary terms actually would be less than before.

[7] U.S. Department of Commerce, *op. cit.*

While it has not been possible to resist making this comparison, I am glad to leave to the reader the decision as to what conclusions if any are to be drawn from it. For obvious reasons, the extent of surplus value taken by itself does not necessarily indicate for either system the share of working class consumption in the national product.

(iv) *Relation of the Government Budget to the National Economic Accounts.* The question is often raised as to how much of the Soviet national income passes through the government budget. The answer for the year 1937, according to our calculations, is 37 percent. (This, to be precise, is the relation of the total government revenue, 107.0 billion rubles, to the gross national product, 291.8 billion rubles.) At first sight, this may seem to be a small figure: in view of the fact that we are dealing with a state in which most economic activity has been socialized, it might be supposed that practically the whole national income would pass through the government budget. The observed relation is understandable, however, if account is taken of: (i) the fundamental Soviet administrative principle already mentioned, according to which the lower echelon agencies responsible for state enterprise are divorced from the government budget and operate on a more or less independent financial basis; (ii) the fact that agriculture is administered largely by collective farms, which also operate on an independent financial basis; (iii) the fact that cooperative enterprise in all spheres also operates on an independent financial basis; and finally (iv) the sizable share of the gross national product that consists of income in kind, which is consumed by the households producing it.

In the United States, in 1937, government revenues totaled $15.4 billions, or 17 percent of the gross national product.

The relation of the various revenue and expenditure items in the budget to the income and outlay categories in our national economic accounts will generally be evident from the captions. In the case of communal services, the government expenditures are only a part of the total outlays for these purposes in the USSR. The balance consists of expenditures by economic enterprises, trade unions, and the social insurance system. The budget item "Financing the National

Economy," amounting to 43.4 billion rubles, includes, according to our calculations, some 8.0 billion rubles of subsidies. At the same time, the balance of 35.4 billion rubles consists almost exclusively of investments of one sort or another. This is 21 billion rubles short of the gross investment for the whole economy. The difference presumably consists of: (i) investments by economic enterprises out of retained profits and depreciation allowances, amounting to about 17 billion rubles in 1937; and (ii) capital investments financed by State Bank credit creation . The State Bank makes some investments in excess of funds made available by the government budget. During 1937, an increase in working capital of economic organizations amounting to nearly 7 billion rubles was financed by bank credit creation.

### RESOURCE ALLOCATION

For purposes of comparison with the data compiled here on the disposition of the Soviet national product by use, Table 8, p. 75, there are available the U.S. Department of Commerce data in Table 11 on the disposition of the national product of the United States in 1937. Unfortunately for our purposes the Department of Commerce does not break down the government expenditures as between different use categories. It is indicated, however, that "New public construction activity" amounted to only 2.0 billion dollars in 1937; so it can be surmised that the total investments by government agencies were of the order of a few billion dollars. If so, the *gross* investment, public and private, would have amounted to about 15 percent of the *gross* national product; and *net* investment, public and private (gross investment less the capital consumption allowances) to about 5 percent of the *net* national product.

According to the data compiled here, then, the Russians in 1937 invested a good deal more and consumed less of their national income than we did of ours in the same year. According to our calculation in adjusted rubles for the USSR, the proportion of the gross national product going to household consumption amounted to 56 percent and to investment to 23 percent. According to the Commerce data the comparable proportions for the United States were respectively 74 percent and (under the assumption stated above) about 15 percent.[8]

So far as consumption is concerned, the foregoing comparisons necessarily are somewhat faulty insofar as the data cited are affected by the difference in the status of such goods as health care and education in the two countries. While "household consumption" for the USSR includes no such goods, since they are (or as of 1937 were) entirely free in that country, the corresponding "personal consumption expenditures" for the United States includes these goods in such amounts as were paid for in this country.

---

TABLE II

NATIONAL INCOME BY USE, UNITED STATES 1937 [a]

|  | BILLION DOLLARS | PERCENT OF GROSS NATIONAL PRODUCT | PERCENT OF NET NATIONAL PRODUCT |
|---|---|---|---|
| a. Personal consumption expenditures | 67.1 | 74.4 | 81.6 |
| b. Government purchases of goods and services | 11.6 | 12.9 | 14.1 |
| c. Gross private investment | 11.5 | 12.8 | 14.0 |
| d. Gross national product | 90.2 | 100.0 |  |
| e. Less: capital consumption allowances | 8.0 |  | 9.7 |
| f. Net national product | 82.2 |  | 100.0 |

[a] "National Income," supplement to *Survey of Current Business*, July, 1947, pp. 19–20. As tabulated here, the gross private investment includes net foreign investment amounting to $.062 billions.

---

Data are not at hand to make the indicated comparison between total consumption, personal and communal, in the two countries. It will be noted, however, that total consumption, including communal services, comprises a smaller share of the Soviet than does personal consumption alone of the American gross national product: 66 percent (using the adjusted ruble data) as compared with 74 percent. The inclusion of free goods available to households in the United

[8] The same relationships obtain if the data of Clark and Wyler are used in place of ours for the USSR; for reasons already stated, however, their calculations lead to lower figures for investment than ours do.

States presumably would raise the American figure by several per-cent.[9]

The foregoing results, then, confirm the prevalent view that by capitalist standards the Russians under their five year plans gave a notably heavy weight to capital investments: it must be observed, however, that the comparisons made may be somewhat misleading in this respect. At any rate, Russian investment policy is placed in a somewhat different light if account is taken of several pertinent facts:

(i) Russian national income in 1937 was far below that of the United States in the same year. According to the data compiled by Clark and Wyler,[10] the net national product per capita in 1937 was $175 to $250 in the USSR and about $640 in the United States (all figures are in terms of United States prices of the year 1937).

(ii) According to the data on United States national income compiled by Simon Kuznets, one must go back to the decade 1869–78 and perhaps earlier to find a period when the American national income was comparable to that of the USSR in 1937. For the decade 1869–78, average yearly net national product per capita of the United States amounted to about $195 (in terms of 1937 prices).[11]

(iii) Contrary to a once prevalent view, the share of the United States national income going to capital formation has not increased in the course of time with the rise in the national income; rather, according to Kuznets' calculations, the ratio was comparatively stable over the entire period 1869–1928, and then declined in the thirties. In the decade 1869–78, gross investment amounted to 18.9 percent of the United States gross national product [12] (in terms of current United States prices).

[9] With regard to the communal services, by the way, one hears widely conflicting statements as to their magnitude in the USSR. The calculations presented in this study provide a basis for judging this question. According to Table 8, the aggregate expenditures on communal services in the USSR in 1937 amounted to 27.4 billion rubles, or 9.4 percent of the gross national product in terms of prevailing prices, and 23.8 billion rubles, or 10.6 percent of the gross national product in terms of adjusted rubles.

[10] See above, p. 86, note 49.

[11] National Product since 1869, pp. 107, 119. Kuznets' data in terms of 1929 prices have been translated into 1937 prices by use of the U.S. Bureau of Labor Statistics index of wholesale prices.

[12] It usually is assumed that the long-term stability in the propensity to save in the United States reflects the operation of a trend factor, but the nature of this factor still

According to these calculations, then, the Russians managed to invest under their five year plans only about 4 percent more of their national product than we did at the comparable historical period.

Needless to say the conclusion is rather surprising, and one immediately wonders at the reliability of the data. As the reader will recall, the calculations of the previous chapter already suggested a limited understatement in investment in terms of adjusted rubles, and obviously one must not take a comparison of the sort attempted very seriously in any event.[13] On the other hand, the calculation still appears to call in question some current notions as to the unprecedented magnitude of Soviet investments under the five year plans.

But from some points of view what is in point evidently is the comparative shares not of investment alone but of investment and defense expenditures together. If the comparison is made in these terms the gap between the two countries is appreciably increased. In addition to their investments, amounting to 23 percent of the gross national product in 1937, the Russians spent on defense another 7.7 percent of the gross national product in the same year. The United States expenditures on defense in the decade 1869–78, by contrast, amounted to only about one percent of the gross national product.

Furthermore, account has to be taken too of differences in the allocation of investments. Presumably the Russians invested more

seems to be conjectural. I wonder if, in this connection, sufficient attention has been given to the rural-urban population shift. It seems to be a well-established fact that the propensity to save is much higher in rural than in urban localities. To this extent, the extensive shift of population from farm to city since the middle of the last century would have operated continually to limit the rise in the average propensity to save for the country as a whole.

[13] One wonders, too, whether there may not be offsetting differences as between the two countries in resource allocation in physical terms, on the one hand, and in the ruble and dollar price structure, on the other. Interestingly, however, the computed share of investment in the gross national product of the United States for the period 1869–78 is slightly larger if 1929 dollar prices are used instead of current prices, i.e., 22.0 percent as compared with 18.9 percent (Kuznets, op. cit., p. 119). There are conceptual differences between the capital investment and national product concepts used in this study and those of Kuznets, particularly in regard to the treatment of government activities, but it is believed these must be of very minor importance in the present comparison. It should be noted that the data of Kuznets cited in these pages refer in all cases to his "peace-time concepts" of the national product.

in basic industries and less in consumers' goods, especially housing, than the United States.

### INDUSTRIAL STRUCTURE

The data in Table 12 for the USSR are from Table 9, columns 7 and 8. Those for the United States are again from the calculations of the U.S. Department of Commerce.

### TABLE 12
### NATIONAL INCOME BY INDUSTRIAL ORIGIN, USSR AND UNITED STATES, 1937

|                                      | NET NATIONAL PRODUCT IN ADJUSTED RUBLES, USSR | | NATIONAL INCOME, U.S. [a] | |
|--------------------------------------|:-------------:|:-------:|:---------------:|:-------:|
|                                      | Billion rubles | Percent | Billion dollars | Percent |
| (1)                                  | (2)           | (3)     | (4)             | (5)     |
| 1. Agriculture                       | 65.2          | 29.9    | 7.20            | 9.8     |
| 2. Industry and construction         | 76.1          | 34.9    | 24.98           | 33.9    |
| 3. Transportation, communications    | 16.8          | 7.7     | 5.58            | 7.6     |
| 4. Trade, including restaurants      | 10.4          | 4.8     | 11.94           | 16.2    |
| 5. Finance                           | 1.9           | .9      | 7.94            | 10.8    |
| 6. Services, including government    | 38.0          | 17.4    | 15.85           | 21.5    |
| 7. Other                             | 1.1           | .5      | .15             | .2      |
| 8. Statistical discrepancy           | 8.7           | 4.0     |                 |         |
| 9. All sectors                       | 218.1         | 100.0   | 73.63           | 100.0   |

[a] U.S. Department of Commerce, "National Income," supplement to *Survey of Current Business,* July, 1947, p. 26. The tabulation shown here is obtained from that in the original source under the following assumptions as to the relation of the categories studied:

| Category in present tabulation | U.S. Department of Commerce categories assumed to be equivalent |
|---|---|
| Agriculture | Farms, agricultural, and similar service establishments |
| Industry and construction | Forestry, fisheries, mining, contract construction, manufacturing, utilities: electric and gas |
| Transportation and communications | Transportation, communication and public utilities, *less* utilities: electric and gas |
| Trade, including restaurants | Wholesale and retail trade |
| Finance | Finance, insurance and real estate, *less* real estate |
| Services, including government | Services, government and government enterprises, real estate |
| Other | Rest of world |

In respect to the scope of the different economic sectors considered, it is believed that, as presented here, these data are roughly comparable to those for the USSR. The American data, however, refer to "national income" in the Department of Commerce sense, which conceptually is almost but not quite the same thing as the Soviet net national product less turnover taxes plus subsidies. (See above, p. 23.)

But making due allowance for this factor, the statistics indicate a notable dissimilarity in the structure of the two economies, and in these terms perhaps a greater backwardness on the part of the Soviet economy than is commonly assumed. At the time considered, when the USSR had experienced about a decade of high-tempo industrialization, the shares of industry and construction and of transportation and communications in the total national product of the two countries were about the same. In the United States, however, industry and construction was by far the dominant sector, while in the USSR the share of agriculture was nearly as great as that of industry and construction. At the same time, trade, finance, and services were all of relatively much less importance in the USSR than in the United States.

But for obvious reasons these differences in structure need not all point to Soviet backwardness. At any rate, account must be taken of the Soviet economists' argument that the comparatively limited development of tertiary production in their country is in some measure an indication not of retarded development but of the avoidance of capitalist waste, e.g., in advertising.

# APPENDICES

# APPENDIX A. NOTES TO TABLE I

## A. INCOMES

*1a. Wages of farm labor.* According to Gosplan, the total wages paid out by agriculture in 1937 amounted to 5.27 billion rubles. On this matter, see *Tretii piatiletnii plan razvitiia narodnogo khoziaistva Soiuza SSR, 1938–42* (Third Five Year Plan of Development of the National Economy of the USSR), p. 229. This Gosplan figure presumably includes wages paid to workers employed on state farms and in machine-tractor stations, but it is believed that it does not include wages paid to hired labor by collective farms. The total wages paid by collective farms to hired labor in 1937 have been estimated at .30 billion rubles. See N.S. Margolin, *Voprosy balansa denezhnykh dokhodov i raskhodov naseleniia* (Problems of the Balance of Money Incomes and Expenditures of the Population), p. 77.

*1b. Money payments to collective farmers on a labor-day basis, etc.* This is the sum of:

(i) The share of the money income of collective farms distributed to collective farm members on a labor-day basis estimated at 6.8 billion rubles (Margolin, p. 7).

(ii) The total "administrative-economic expenses" of collective farms, .26 billion rubles, which is assumed to consist mainly of salaries of officials and administrative personnel of the collective farms. This figure is calculated from the following data. The total money income of collective farms in 1937 amounted to 14.18 billion rubles: TSUNKHU, "Sotsialisticheskoe sel'skoe khoziaistvo Soiuza SSR" (Socialist Agriculture of the USSR), as reprinted in *Planovoe khoziaistvo* (Planned Economy), 1939, No. 7, p. 144; the "administrative-economic expenses" of collective farms amounted to 1.8 percent of the total money income: M. Nesmii, "Dokhody kolkhozov i kolkhoznikov" (Incomes of the Collective Farms and Farmers), *Planovoe khoziaistvo*, 1938, No. 9, p. 98.

(iii) The money premiums paid collective farm members, estimated at .28 billion rubles. According to Nesmii (*op. cit.*, p. 98), 3.9 percent of the total money income of collective farms was allocated to funds for social and cultural measures, premium payments, etc. I assume that one half of this sum actually was paid out in the form of premiums, the other half being devoted to social and cultural measures, and other purposes.

*1c. Net money income from the sale of farm products.* This is calculated as the sum of: (i) the sales of individual collective farmers (as distinct from collective farms as such) and of other farm households on the retail collective farm market, estimated at 13.68 billion rubles; and (ii) the sales of individual collective farmers (as distinct from collective farms as such) and of other farm households to government and other procurement agencies, estimated at 2.88 billion rubles; *less* (iii) an allowance for the money expenses of the farm households (for seed, etc.), 2.36 billion rubles.

The estimates on the first two items are obtained as follows:

According to TSUNKHU ("Sotsialisticheskoe sel'skoe khoziaistvo," p. 144), total sales by the collective farms both to the government procurement agencies and on the collective farm market amounted to 12.00 billion rubles in 1937. (According to the Soviet source, this figure of 12.00 billion rubles represents the total money income of collective farms from "crops" and "livestock," while the total money income from all sources was 14.18 billion rubles. The difference of 2.18 billion rubles represents, it is believed, the earnings of collective farms from nonfarm activities—flour and vegetable oil mills, smithies, etc.). On the basis of a breakdown of the corresponding figure for 1936, given in Nesmii ("Dokhody kolkhozov i kolkhoznikov," p. 95), it is estimated that of the total sales—12.00 billion rubles—the sales to government procurement agencies amounted to 65.7 percent or 7.88 billion rubles and those on the collective farm market to 34.3 percent or 4.12 billion rubles. According to Margolin (*Voprosy balansa denezhnykh dokhodov i raskhodov naseleniia*, p. 63), the total sales on the collective farm market from all sources amounted to 17.8 billion rubles in 1937. From this total I deduct 4.12 to allow for sales by collective farms (see above). It is assumed that the balance of the sales, or 13.68 billion rubles, represents the sales by collective farm and other households. (A further deduction should be made, however, to allow for an unknown but probably small amount of sales by state farms.)

The sales by collective farms as such to government procurement agencies are calculated at 7.88 billion rubles. As a first approximation, the sales of individual collective and other farm households to these same agencies are taken to be in the same proportion to the sales by collective farms as the total output of these farm households is to the total output of collective farms. The data on the output of different types of farm organizations is given in TSUNKHU, "Sotsialisticheskoe sel'skoe khoziaistvo," p. 160. It is calculated that the total output of the farm households amounts to 36.6 percent of the total output of the collective farms; hence the sales of the farm households to the procurement agencies are taken to be 2.88 billion rubles.

The figure on the money expenses of farm households is a more or less arbitrary estimate obtained as follows: On the basis of scattered information including data in Nesmii (*op. cit.,* pp. 102–103), it is assumed that the money expenses amount to 10 percent of the total money incomes of the farm households. The money incomes comprise:

|  | BILLION RUBLES |
|---|---|
| Payments to collective farms on a labor-day basis | 6.80 |
| Premiums paid to collective farmers | .28 |
| Sales by farm households on collective farm markets | 13.68 |
| Sales to government and cooperative procurement agencies | 2.88 |
| Total | 23.64 |

*1d. Net farm income in kind.* Net farm income in kind, as valued in terms of average prices realized on the farm, is calculated here as follows:

(i) It is estimated that of the gross harvest of all agricultural products, 36.15 percent was marketed in 1937. This estimate is based on the following information: data on the amounts in physical units of different agricultural products marketed in 1937, from TSUNKHU, "Sotsialisticheskoe sel'skoe khoziaistvo," p. 162; data on the 1937 gross output of these various products in physical units from *ibid.,* p. 153 and Gosplan, *Tretii piatiletnii plan,* pp. 218 ff; and data on the value in terms of 1926–27 prices of the total output of different branches of agriculture as of 1937, from TSUNKHU, "Sotsialisticheskoe sel'skoe khoziaistvo," pp. 160–61 and I. V. Sautin, *Kolkhozy vo vtoroi Stalinskoi piatiletke* (Collective Farms in the Second Stalinist Five Year Plan), pp. xi and 81.

(ii) The total sales on the retail collective farm market from all sources, as indicated above in item 1c, amounted to 17.8 billion rubles in 1937. According to Gosplan (*Tretii piatiletnii plan,* p. 88), the total value of sales to procurement agencies from all sources amounted to 15.7 billion rubles in 1937. This is taken to be the value of procurements at procurement prices and is assumed to include deliveries to the MTS which, it is believed, are valued for bookkeeping purposes at these same prices. Finally, as is explained subsequently, the government subsidized agriculture, particularly the MTS and state farms, to the extent of 4.5 billion rubles in 1937 (Appendix C, pp. 128 ff.).

(iii) The sum of the collective farm sales and the value of procurements, gross of subsidies, 38.0 billion rubles, is taken to indicate the value of the marketed share in terms of average current prices realized by the farmer.

(iv) On the basis of (i) and (iii), the value of the gross harvest in 1937, in terms of these same prices, may be estimated at 105.11 billion rubles (i.e., $38.0 \div 36.15 \times 100$).

(v) From this sum, we deduct 3 percent or 3.15 billion rubles for a reason to be stated in a moment. This leaves a balance of 101.96 billion rubles, which may be referred to as the adjusted gross harvest. On the basis of a variety of scattered information, it is assumed that in reporting their grain harvest the Russians in 1933 changed from a "barn yield" basis (i.e., the amount of grain actually harvested) to a "biological yield" basis (i.e., the amount of grain in the field before harvesting), less an allowance of some 10 percent for harvesting and threshing losses; and that in 1937 they changed again to a full biological yield. In our calculation, the deduction of 3 percent of the gross harvest is intended to put the harvest figure for 1937 on a basis comparable to the figures for immediately preceding years. It happens that the grain harvest in 1937 was about one third of the gross harvest of all agricultural products (see TSUNKHU, "Sotsialisticheskoe sel'skoe khoziaistvo," p. 160), so that a 3 percent reduction in the gross harvest of all agricultural produce corresponds roughly to a 10 percent reduction in the grain crop.

(vi) From the adjusted gross harvest we deduct 35 percent, or 35.69 billion rubles, to allow for production expenses in money and in kind; this leaves a calculated net harvest of 66.27 billion rubles. According to data in Central Administration of Economic and Social Statistics, *Socialist Construction in the USSR*, pp. 30, 238, the ratio of production expenses to the gross production of agriculture amounted in 1933 to 34 percent; in 1934 to 36 percent; and in 1935 to 40 percent. In view of the unusually large harvest in 1937, presumably the ratio of production expenses to the volume of the harvest would have been less than the average for these years. The allowance of 35 percent for 1937 is made with this fact in mind. Insofar as there are field to barn losses in grain, over and above the standard 10 percent allowance made in the crop reports for 1933–35 (it is believed that the losses in fact appreciably exceed 10 percent), our calculation assumes that these losses are allowed for in production expenses. Soviet sources are not clear on this point.

(vii) Deducting the marketed share as computed above, we find that the nonmarketed share of the net harvest amounted to 28.27 billion rubles.

(viii) To this figure, we add an allowance for net money production expenses incurred outside agriculture, 6.30 billion rubles. The result, 34.57 billion rubles, is taken to be net farm income in kind. In order to obtain an estimate of income in kind, these money expenses should not have been deducted from the gross harvest in the first place; insofar as they are deducted in (vi) above, this is now canceled by the addition of the corresponding sum. The expenses in question comprise chiefly expenses for fuel, lubricants, repairs, and fertilizer, though some seed also is purchased by agricultural from non-agricultural enterprises. The figure of 6.30 billion rubles is a rough estimate of the total of these expenses and is based on scattered information, including chiefly data in N. N. Rovinskii, *Gosudarstvennyi biudzhet Soiuza SSR* (Government Budget of the USSR), p. 197, and M. Nesmii, "Finansovoe khoziaistvo kolkhozov" (Financial Economy of the Collective Farms), *Planovoe khoziaistvo*, 1939, No. 8, pp. 100, 103.

(ix) As calculated above, net farm income in kind includes investments out of income in kind by farm organizations, which are taken to amount to some 2.00 billion rubles. In order to obtain a figure representing the net farm income in kind accruing to households it is necessary to deduct the amount of these investments. This has been done here, and the result rounded to 32.50 billion rubles to avoid an undue appearance of precision. The estimate of investments out of income in kind by farm organizations, which is included in Table 2, is explained below, Appendix B, p. 112. For the purposes of Table 2, it should be noted that this estimate is taken to refer to the increase in capital of all sorts, including in particular seed and fodder funds, livestock herds, unfinished production in the form of ploughed land, etc., and labor investments in farm buildings, etc. According to the logic of the present calculation, however, the gross harvest figure derived in (iv) above has the same scope as Soviet official statistics on agricultural output, and it would seem that not all of the indicated forms of investment are included in the Soviet figures: While seed and fodder funds and unfinished production are so included, labor investments probably are not, and in the case of livestock herds the Soviet treatment is not entirely clear.

On this whole question, see D. I. Chernomordik, ed., *Narodnyi dokhod SSSR* (National Income of the USSR), pp. 94 ff.; and Paul Studenski, "Methods of Estimating National Income in Soviet Russia," *Studies in Income and Wealth,* Vol. VIII, pp. 208 ff. On the basis of information in Chernomordik, pp. 103 ff., Studenski concludes that in the official data on agricultural output, no allowance is made for changes in livestock herds. From Chernomordik's discussion, however, the present writer is led to believe that while this may have been so before 1936, such an allowance was made beginning in that year. Also, there is a question in my mind whether Chernomordik, himself, has correctly interpreted the official data for years prior to 1936. An explanation of the official statistics, quoted by Chernomordik, p. 103, from Gosplan's instructions might be construed as providing implicitly for an allowance for changes in livestock numbers.

Insofar as one or another type of investment is not included in the Soviet data on gross agricultural production, no deduction for such investments should be made here.

It should be noted finally that the estimate of income in kind derived in Table 2 refers to *collective farm* investments only, and does not include investments by *state farms*. Strictly speaking, an allowance ought to be made here also for state farm investments of this sort. Because of the limited role of these farms, however, such investments must have been of quite small magnitude.

2. *Wages and salaries, nonfarm.* According to Gosplan (*Tretii piatiletnii plan,* pp. 228–29), the wage bill of the USSR in 1937 amounted to 82.25 billion rubles. This sum includes 5.27 billion rubles of wages paid to agricultural workers. Our figure on the amount of nonfarm wages and salaries recorded in the statistical current reports, 76.98 billion rubles, is obtained by subtracting from the indicated wage bill the indicated amount paid to agricultural workers.

While this figure on the wage bill, which was compiled by the Soviet statistical agency TSUNKHU, usually is interpreted as a comprehensive figure for the USSR, this is not the case; on the basis of information in Abram Bergson, "A Problem in Soviet Statistics," *Review of Economic Statistics,* November, 1947, it is calculated that the TSUNKHU figure of 82.25 billions for 1937 actually is only 78.2 percent of the total wage bill of the USSR in that year. (This figure is obtained by interpolation from corresponding figures for 1934 and 1938—77.2 and 78.5 percent—which are believed to be the most reliable of the pertinent data presented in the article cited.) The total wage bill, then, comes to 105.18 billion rubles.

The gap between the TSUNKHU and the comprehensive total for the wage bill is accounted for to some extent, it is believed, by certain premium funds, probably to a greater extent by wages paid workers in certain local industries and in secondary lines of activity in certain industries; possibly also to a considerable extent by the wages paid convict labor; and finally, very likely to some extent by army pay (see, on this whole question, Bergson, *ibid.;* and Harry Schwartz, "A Critique of 'Appraisals of Russian Economic Statistics,'" *Review of Economic Statistics,* February, 1948).

The figure in Table 1, A, item 2b, "Wages and salaries, nonfarm, other," 21.13 billion rubles, represents the gap between the TSUNKHU and the comprehensive figure on the wage bill, 22.93 billion rubles, less two deductions: (i) the item of .30 billion rubles wages that collective farms pay hired labor, which is included in the comprehensive but not in the TSUNKHU figure; (ii) an estimated 1.50 billion rubles of army pay. The calculation of the latter figure is explained below.

3. *Income of artisans; other money income, currently earned.* This is calculated chiefly on the basis of data on the "earnings of cooperative artisans" and "other income" reported in N. S. Margolin, *Balans denezhnykh dokhodov i raskhodov naseleniia* (Balance of the Monetary Incomes and Expenditures of the Population), p. 8. Margolin presents here percentage figures on the share of these and other items, including the wage bill, in total money income in 1934 and 1938. On the basis of absolute figures on the wage bill for 1934 and 1938, presented in Bergson, *op. cit.*, the "earnings of cooperative artisans" and "other income" in 1934 and 1938 can be calculated in absolute terms. Corresponding figures for 1937, totaling 16.30 billion rubles for the two items together, were obtained by interpolation.

From this total we deduct 2.60 billion rubles, which is taken to indicate the total amount of interest receipts on government bonds and savings deposits together with repayments of principal. This is a planned figure, taken from M. Bogolepov, "Finansy SSSR nakanune tret'ei piatiletki" (Finances of the USSR on the Eve of the Third Five Year Plan), *Planovoe khoziaistvo*, 1937, No. 3, p. 112. Margolin includes these payments in "other income"; we deduct them because they are treated here in this study as transfer payments rather than as income currently earned. Margolin's category "other income" may include certain other receipts which are not currently earned (e.g., receipts from sales of second hand clothing); if allowance were made for these items, the estimated total of "income of artisans" and "other money income currently earned" would be reduced still further.

The category "other income" is believed to include, however, a number of items which would represent income currently earned: the earnings of independent artisans, domestic workers, haulers, payments to gold miners for their gold, etc.

4. *Imputed net rent of owner-occupied dwellings.* On the basis of scattered information, it is assumed (i) that the total amount of floor space privately owned and occupied in the USSR (in both the country and the city) amounted to about 450 million square meters in 1937; and (ii) that this floor space, at the average rental paid by urban workers for rented dwellings, would have a value of about 75 kopeks per square meter per month or 9 rubles per square meter per year.

5a. *Military pay.* On the basis of scattered information, including data in S. N. Kournakoff, *Russia's Fighting Forces,* p. 65, the armed forces of 1937 are taken to number 1,750,000 men, including 175,000 commissioned officers. It is assumed that the average earnings of the privates and non-coms taken together amounted to 150 rubles a year and of the officers, 8,000 rubles a year. This assumption is based on a consideration of information in (i) *Postanovlenie Soveta narodnykh komissarov* (Decree of the Council of People's Commissars), November 5, 1932, *O povyshenii okladov soderzhaniia lichnomu sostavu Raboche-krest'ianskoi krasnoi armii* (On the Increase of Pay of the Personnel of the Workers' and Peasants' Red Army), *Sobranie zakonov i rasporiazhenii SSSR* (Collected Laws and Decrees of the USSR), November 15, 1932; in (ii), *The Land of Socialism Today and Tomorrow* (Reports and Speeches at the Eighteenth Congress of the Communist Party . . . March 10–21, 1939), Moscow, 1939, p. 288; and in (iii) W. H. Hutt, "Two Studies in the Statistics of Russia," *The South African Journal of Economics,* March, 1945. On the basis of these assumptions, the total military pay comes to 1.66 billion rubles; in order to avoid the appearance of precision, this sum has been rounded to 1.50 billion rubles.

*5b. Military subsistence.* It is assumed more or less arbitrarily that the value, at prices paid by the defense commissariats, of the food and clothing provided for the armed forces amounted to about 1,500 rubles per man in 1937. According to TSUNKHU, *Trud v SSSR* (Labor in the USSR), p. 343, the average annual expenditure of a working class family on food and clothing amounted to 741 rubles per head in 1935.

*6. Statistical discrepancy.* This is calculated as the difference between the sum of incomes, both earned and unearned, as recorded in Table 1, A, and the sum of outlays, as recorded in Table 1, B.

*8. Pensions and allowances.* Margolin, *Voprosy balansa denezhnykh dokhodov i raskhodov naseleniia,* pp. 84, 85.

*9. Stipends.* The sources, basic data, and procedures used to obtain this estimate of 2.16 billion rubles, which refers to allowances to students in higher educational institutions, are the same as are described above, in respect of the calculation of the "earnings of cooperative artisans" and "other income" (item 3).

*10. Interest receipts.* Estimated at .90 billion rubles on the basis of data on the outstanding debt and savings deposits held by individuals, in A. Baykov, *The Development of the Soviet Economic System,* p. 380. Baykov's tabulations are taken in turn from V. P. D'iachenko, ed., *Finansy i kredit SSSR* (Finances and Credit of the USSR), pp. 264–268. It is assumed that the interest on government bonds averaged 4 percent and on savings deposits, 3 percent in 1937 (see A. Z. Arnold, *Banks, Credit and Money in Soviet Russia,* pp. 500–501, 504n.).

### B. OUTLAYS

*1. Retail sales to households.* According to Gosplan (*Tretii piatiletnii plan*, p. 233), the total retail sales by government and cooperative shops and restaurants amounted to 125.94 billion rubles in 1937. According to Margolin, *Voprosy balansa denezhnykh dokhodov i raskhodov naseleniia*, p. 8, sales on the collective farm market amounted to 17.8 billion rubles in the same year. These totals include sales not only to households but also to institutions. Sales to institutions are estimated to have amounted to 11.5 percent of the total turnover of government and cooperative shops and to 10 percent of the turnover in the collective farm markets (based on data in *ibid.*, pp. 63, 101).

*2. Housing (including imputed net rent of owner-occupied dwellings); services.* According to Margolin, *Balans denezhnykh dokhodov i raskhodov naseleniia*, p. 9, the total money outlays for rent, utilities, amusements, passenger transportation, and other services amounted to 10.5 percent of all consumers' money outlays in 1934 and to 9.6 percent of all consumers' money outlays in 1938. Margolin also presents corresponding percentage figures on money outlays for goods, including expenditures for meals in restaurants; so that it may be calculated that the money outlays for housing and other services were 13.6 percent of the outlays for goods in 1934 and 12.1 percent of the outlays for goods in 1938. By interpolation, it is computed that the money outlays for housing and other services amounted to 12.5 percent of the outlays for goods in 1937. Using the figure already calculated on the outlays for goods in this year in absolute terms, the sum of the money outlays on housing and services in 1937 is computed at 15.94 billion rubles.

To the money outlays for housing and other services is added 4.00 billion rubles to allow for the imputed net rent of owner-occupied dwellings. Actually, the item to add here would be imputed gross rent; it is possible, however, that the figure on money outlays for housing and services includes expenditures for housing maintenance and amortization which would offset the depreciation on owner-occupied dwellings.

*3. Trade union and other dues.* According to V. Cherniavskii and S. Krivetskii, "Pokupatel'nye fondy naseleniia i roznichnyi tovarooborot" (Purchasing Power of the Population and the Retail Turnover), *Planovoe khoziaistvo*, 1936, No. 6, trade union and other dues amounted to .73 billion rubles in 1935. This is about one percent of the total wage bill, including agricultural wages, in that year (Bergson, "A Problem in Soviet Statistics," p. 236). It is assumed that this relation obtained in 1937.

*4. Consumption of farm income in kind: army subsistence.* See items on net farm income in kind, and on military subsistence.

*6a. Net bond purchases.* Calculated from data in Baykov, *The Development of the Soviet Economic System,* p. 380.

*6b. Increment of savings deposits.* Calculated from data in Baykov, p. 380.

*6c. Other, including increment in cash holdings.* In regard to cash, we allow 1.0 billion rubles for the increment of holdings in 1937. Here we are guided chiefly by data in Arnold, *Banks Credit and Money in Soviet Russia,* p. 412, which indicate that during the period 1931–35 the total currency in circulation increased at an average rate of .8 billion rubles per annum. There is no reason to think that the increment in cash holdings in 1937 would have been substantially in excess of this amount. Besides cash, the other forms of savings are initiation payments to cooperatives, life insurance, etc. Our allowance of .5 billion rubles for these items is more or less arbitrary.

*7. Direct taxes.* Narkomfin, *Gosudarstvennyi biudzhet SSSR za vtoruiu piatiletku, 1933–37 gg.* (State Budget of the USSR for the Second Five Year Plan 1933–37), pp. 8–11.

# APPENDIX B. NOTES TO TABLE 2

## A. INCOMES

*1. Net income retained by economic organizations.*

*1a. Retained income in kind of collective farms.* This is a somewhat arbitrary estimate of the increment of collective farm capital of all sorts—e.g., seed and fodder, livestock, work in process and construction—that was financed out of income in kind as distinct from cash purchases. According to M. Nesmii, "Finansovoe khoziaistvo kolkhozov" (Financial Economy of the Collective Farms), *Planovoe khoziaistvo,* 1939, No. 8, p. 95, the increase in seed and fodder funds and livestock herds and the volume of labor investments in construction in Soviet collective farms averaged 95.3 rubles per household in 1937. Seed and fodder and livestock are valued here at government procurement prices; it is not indicated how labor investments are valued. While Nesmii expresses investments in per household terms, it seems clear from the context that he is referring to collective farm investments only and not to investments by collective farm households on their own homesteads. Since there were 18 million such households in 1937, the aggregate investments of the specified sort come to 1.7 billion rubles. Our estimate of collective farm investments out of income in kind, 2.00 billion rubles, represents a global estimate, which is intended to take into account the following facts: (i) The figure of 1.7 billion rubles on collective farm investment just derived from Nesmii's data. (ii) The fact that income in kind in the present study is valued at the average of all realized farm prices, including retail collective farm market prices, while investment in kind in Nesmii is valued at the average of procurement prices only. (iii) Investments in work in process, i.e., ploughing, etc., are not included in Nesmii's figure; it is assumed that ours takes them into account. (iv) It is not clear whether and to what extent the collective farm investments itemized by Nesmii include investments made in the form of cash purchases as distinct from investments out of income in kind; cash purchases presumably should *not* be included in collective farm investments for our present purposes.

*1b. Retained money income of collective farms.* This is calculated as the sum of: (i) Money income allocated to funds for capital investment and debt repayment, amounting to 1.76 billion rubles (see Nesmii, in *Planovoe khoziaistvo,* 1938, No. 9, p. 98; TSUNKHU, "Sotsialisticheskoe sel'skoe khoziaistvo, p. 144). (ii) One half the money allocations to funds for premiums and social-cultural measures, or .28 billion rubles (see Appendix A, p. 103).

*1c. Net income retained by state and cooperative enterprises.* According to Gosplan (*Tretii piatiletnii plan,* p. 115), the total profits of state and cooperative enterprises amounted to 17 billion rubles in 1937. According to Narkomfin (*Gosudarstvennyi biudzhet SSSR za vtoruiu piatiletku, 1933–37 gg.,* pp.

8–11), the total payments out of the profits of enterprise (*otchisleniia ot pribylei*) into the government budget, amounted to 9.29 billion rubles. These payments do not include the income taxes paid into the budget by cooperative enterprise which, according to Narkomfin, pp. 8–11, amounted to .75 billion rubles. Hence the profits retained by state and cooperative enterprise come to 6.96 billion rubles.

While Soviet figures on profits of the sort cited here usually are interpreted as true net profits figures, and the one cited here was so interpreted in an earlier version of this study, it is now believed that the figure of 17.0 billion rubles represents instead total profits prior to any allowance for subsidized losses. This interpretation is based on a detailed analysis, undertaken jointly by the writer and Hans Heymann, Jr., of Soviet practice in accounting for subsidized losses and compiling data on profits. It is hoped that a memorandum embodying this analysis will be made generally available at an early date.

Also, the retained profits as calculated here represent profits before allocations to the so-called Director's Fund, established in April, 1936. Since premiums paid to workers out of this fund supposedly are already included in wages, this means that there is a slight amount of double-counting in our calculation of the gross national product.

2. *Charges to economic organizations for special funds.*

2a. *For social insurance.* See Narkomfin (pp. 8–11, 77); K. N. Plotnikov, *Biudzhet Sovetskogo gosudarstva* (Budget of the Soviet Government), p. 75.

2b. *For trade unions, etc.* This comprises: (i) Expense charges for contributions to the trade unions' budget. These contributions, which together with dues constitute the principal revenue of the trade unions and the amounts of which are fixed by law, are taken to be 1.45 percent of the total wage bill for 1937, as reported by TSUNKHU, or 1.19 billion rubles. According to the financial plan for 1936, the contributions were to total 1.04 billion rubles in 1936, or 1.45 percent of the wage bill reported by TSUNKHU for that year (A. Lozovsky, *Handbook on the Soviet Trade Unions*, p. 28; Baykov, *The Development of the Soviet Economic System*, p. 342). (ii) Expense charges for contributions to special funds for workers' training and education, including the operation of factory apprentice schools and of courses for more advanced training. These contributions, which likewise are fixed by law, are taken to be about one percent of the total wage bill for the whole economy, or 1.00 billion rubles for the year 1937. See M. V. Nikolaev, *Bukhgalterskii uchët* (Accounting), pp. 421 ff., where the amounts of contributions by different commissariats are stated as a percentage of the wage bill. While the percentages cited by Nikolaev average more than one percent, apparently only the funds levied on the wages of workers engaged in production were an expense charge before profits.

As to our assumption that the foregoing contributions generally were charged as expense before profits, see Nikolaev, pp. 121 ff., 421 ff., 564 ff., D. C. Levin and M. G. Poliakov, *Kal'kulirovanie sebestoimosti produktsii miasnoi promyshlennosti* (Calculation of Costs of Production of the Meat Industry), p. 15.

*3a. Taxes on incomes of collective farms.* Narkomfin, pp. 8–11.

*3b. Payments from profits of state and cooperative enterprise to government budget.* See above, item 1c.

*3c. Turnover tax.* Narkomfin, pp. 8–11.

*3d. Miscellaneous.* This consists to the extent indicated of the following items from Narkomfin, pp. 8–11.

|                                                   | BILLION RUBLES |
|---------------------------------------------------|---------------:|
| Tax on state farms                                | .04 |
| Tax on noncommodity operations                    | .36 |
| Local taxes on enterprises of socialized economy  | .46 |
| Customs revenue                                   | 1.32 |
| Rental income from property of local Soviets      | .55 |
| Forestry incomes                                  | .36 |
| Total of specified items                          | 3.09 |

In addition to the foregoing, Narkomfin, pp. 8–11, lists a category of revenue "other incomes" of 3.25 billion rubles. It is arbitrarily assumed here that of this total .91 billion rubles of revenue is from indirect taxes and fees other than those listed above, e.g., fines paid by organizations, auto inspection fees, etc. (See Appendix D.)

It should be noted that some of the charges listed above, e.g., the forestry income (levied in the form of stumpage fees), are collected from households as well as organizations; hence insofar as they have not already been deducted in calculating the net income of the households, the revenue from them should be included with the direct taxes and charges in Table 1 rather than in the indirect taxes and charges in Table 2.

Also with regard to the item "Rental income," it should be noted that this was on a gross rather than net basis since, unlike economic enterprises generally, housing up to October, 1937 did not operate on the so-called system of "economic calculation." The logic of Table 2 would dictate, however, that the *net* income from housing and not the gross revenue be included on the income side. For a discussion of the various types of Soviet government revenue, including those listed above, see A. K. Suchkov, *Dokhody gosudarstvennogo biudzheta SSSR* (Incomes of the State Budget of the USSR).

*4. Allowance for subsidized losses.* It has been indicated that this is to be taken as the sum of the following:

(i) Planned budget appropriation for "outlays on mastering of production" (*raskhody po osvoeniiu proizvodstva*), 6.5 billion rubles;

(ii) Budget outlays to compensate for payments by procurement agencies to farmers for extra-norm deliveries of farm produce, etc., .5 billion rubles;

(iii) Excess of actual over planned outlays "on mastering production," 1.0 billion rubles.

As a part of their system of financial planning, Soviet planners compile data not only on the incomes and expenditures of the government but also on the sources and disposition of Soviet finance generally. Published summary tabulations on the latter are very much like our own consolidated account of the incomes and outlays of government, economic organizations, etc., in Table 2, p. 20. The figure cited above on planned "outlays on mastering of production," it is to be observed first, actually is taken not from the government budget but from such a summary tabulation. More particularly, according to Smilga, "Finansy sotsialisticheskogo gosudarstva" (Finances of the Socialist Government), *Problemy ekonomiki* (Problems of Economics), 1937, No. 2, p. 115, outlays "on mastering" (*po osvoeniiu*) in the plan for finance generally were to amount to 6.5 billion rubles in 1937. But it is believed that these outlays represent at the same time the outlays under the corresponding heading in the government budget. In the case of some outlays in the plan for finance generally, funds are derived not only from the budget but from other sources, e.g., investments in working capital are financed partly out of budget appropriations and partly out of retained profits. But all the indications are that so far as "outlays on mastering of production" are concerned, the plan for finance generally and the budget come to the same thing; accordingly, I have ventured to proceed here on this basis.

At the same time, "outlays on mastering of production" is the established Soviet euphemism for subsidies; and it seems clear that with minor exceptions the outlays in question come to the same thing as subsidies as understood in this study. Unfortunately, by the way, full details on the Soviet budget classification at the time studied are lacking, and there is some question as to just what caption in the budget corresponds to that in the financial plan for 1937. In all probability it was essentially the same as that in the plan. The caption, outlays "on mastering of production" (*po osvoeniiu proizvodstva*) is known to have been used in 1936. But possibly use was made of the more explicit *dotatsiia*. This last caption definitely appears in the budget classification for 1938.

As to the exceptions, there is some doubt also regarding these, but at the time studied probably the most important is the one taken into account in our calculation of the total subsidized losses above, i.e., the appropriations to procurement agencies to compensate for payments made to farmers for deliveries of certain agricultural products in amounts above the obligatory norms and for certain other outlays. These appropriations, it is believed, were recorded not under the budget heading referring to subsidies generally but, along with appropriations for such special measures as industrial resettlement, geological surveys, etc., under another heading, "operational outlays" (*operatsionnye raskhody*).

As to the amount of these payments, I have merely sought here to summarize broadly a variety of scattered information, including chiefly the schedule of payments in "Postanovlenie Soveta narodnykh komissarov . . ." ("Decree of the Council of People's Commissars . . ."), *Finansovoe i khoziaistvennoe zakonodatel'stvo* (Financial and Economic Legislation), 1936, No. 6, pp. 12 ff., and in the later decree cited above, p. 25, note 5; various data on procurements and procurement prices; and budget statistics in Narkomfin, *Otchët ob ispolnenii gosudarstvennogo biudzheta SSSR za 1935* (Account on the Fulfillment of the Government Budget of the USSR for 1935), pp. 58 ff.

To come now to the subsidies in excess of the plan, it may be advisable first to recall the doubts expressed earlier as to the nature of Soviet financial practice on this question. More particularly, losses beyond the initially planned amounts may be financed by bank credit creation instead of by extra subsidies from the budget; just which is done and to what extent is conjectural. But, for present purposes, except where reference is to the government budget, losses financed in one way seem to come to much the same thing as losses financed in the other. Accordingly, our allowance for 1.0 billion rubles of extra subsidies, then, might be viewed as in effect allowing for either of these two kinds of losses. At the same time, this figure is quite arbitrary, being intended to give effect chiefly to the following information:

(i) In the financial plan referred to above, the goal for profits was 20.8 billion rubles (Smilga, p. 110). But profits actually amounted to 17 billion rubles in 1937 (See the item on net income retained by state and cooperative enterprises, p. 113). These figures represent profits gross of the losses of subsidized organizations, and accordingly do not necessarily indicate the developments regarding such losses. But they obviously are suggestive. The implication is that there probably were losses beyond those planned initially and at the same time that these were rather limited.

(ii) In the plan for the consolidated government budget for 1937, the goal for total outlays on the "National Economy" was fixed at 43.0 billion rubles. This was to be constituted as follows:

|  | GOAL BILLION RUBLES |
|---|---|
| a. Investments in fixed capital, other than "extra-limit" outlays | 18.6 |
| b. "Extra-limit" outlays | 1.4 |
| c. Investments in "own" working capital of economic organizations | 6.4 |
| d. Subsidies | 6.5 |
| e. Reserves | 1.7 |
| f. Other outlays | 8.4 |

Here, "extra-limit" investments represent small investments not listed by title in the annual plan approved by the union and republican governments. "Own" working capital represents the working capital permanently assigned to the economic organization to meet minimum needs, as distinct from working capital financed from bank loans to meet seasonal and other special needs. Subsidies are those included in the pertinent Soviet budget category. The residual item is believed to consist mainly of "operational expenditures."

The above computations, except for the figure on reserves, are found or calculated from figures in Smilga, pp. 116–119. The figure on reserves is from G. F. Grin'ko, *Finansovaia programma Soiuza SSR na 1937* (Financial Program of the USSR for 1937), p. 72. In Smilga, budget "extra-limit" outlays are given as 1.6 billion rubles. I have deducted .2 billions to allow for outlays on "social-cultural" measures, which are believed not to be included under outlays on the "National Economy" in the Soviet budget.

These foregoing data represent the goals for 1937. On the actual results, there is at hand the following information:

(a) The total outlays on the "National Economy" amounted to 43.4 billion rubles (Narkomfin, *Gosudarstvennyi biudzhet SSSR za vtoruiu piatiletku*, p. 9), or .4 billions in excess of the plan.

(b) Investments in fixed capital amounted to 18.3 billion rubles (*ibid.,* p. 76). It is not clear whether this is exclusive or inclusive of "extra-limit" outlays. If exclusive, the goal would have been underfulfilled by .3 billion rubles; if inclusive, the goal would have been underfulfilled by 1.7 billion rubles.

(c) In the case of investments in "own" working capital, available information is conflicting but clearly the goal for 1937 was appreciably underfulfilled. This refers, however, to all investments, including not only budget appropriations but also investments out of the retained profits of the economic organizations, etc., and most likely the deficiency was primarily in regard to the latter. The possibility still is open that there may have been a deficiency also in budget appropriations, but this could have been only a limited sum. The information referred to may be tabulated as follows:

| | "OWN" WORKING CAPITAL OF ECONOMIC ORGANIZATIONS, BILLION RUBLES | | |
| --- | --- | --- | --- |
| | Actual Jan. 1, 1937 | Goal Jan. 1, 1938 | Actual Jan. 1, 1938 |
| Smilga, *op. cit.,* pp. 118–119; Grin'ko, *op. cit.,* pp. 38–39 | 37.3 | 48.3 | |
| Bogolepov, *op. cit.,* p. 115 | 42 | 57 | |
| N. Sokolov, "Bor'ba Gosbanka za khoz-raschët v tret'ei piatiletke" (Struggle of the State Bank for Economic Accounting in the Third Five Year Plan), *Planovoe khoziaistvo,* 1939, No. 6, p. 36 | | | 45.7 |
| *Second Session of the Supreme Soviet of the USSR,* August, 1938, p. 536 | | 49.0 | 45.7 |

Very likely the inconsistency as between Bogolepov's data and those from other sources reflects a difference in coverage. More particularly, Bogolepov may be referring to both state and cooperative enterprises while the other sources are referring only to state enterprises. At the same time, it is believed that practically all the budget investments in working capital would be in state enterprises, so the data on these enterprises are the ones of interest here. Referring to the state enterprises, then, the planned increment in working capital was 11 and the realized one 8.4 billion rubles.

In sum, the goal for the total budget outlays on the "National Economy" was slightly overfulfilled, while those for two major components, investments in fixed and working capital, may have been somewhat underfulfilled. This again suggests that the goal for subsidies may have been overfulfilled, but only by a limited amount.

Our estimate for total subsidies is 8.0 billion rubles. At this point reference may be made to a puzzling passage in Gosplan, *Tretii piatiletnii plan . . . ,* p. 114, particularly:

"All profits (*pribyl*) of state enterprises and cooperative systems will amount to 51.3 billion rubles in 1942 compared with 17 billion rubles in 1937 (an increase of three times). This growth reflects the increase in production and lowering of costs, and also the changes in the relation of profits and the turnover tax in the price of goods that occurred in 1938 and 1939.

"The amount of profits (deducting for losses) will amount to 33.3 billion rubles in 1942 compared with 7 billion rubles in 1937. The basic factor in the increase in profits in industry is the lowering of costs.

"The accumulation (*nakopleniia*) of transport and communications will grow from .1 billion rubles in 1937 to 3.1 billion rubles in 1942."

As already has been explained, the Gosplan profits figure of 17 billion rubles for 1937 is interpreted in this study as representing net profits before allowance for losses of subsidized enterprises. The precise meaning of the "losses" (*ubytki*) referred to in the above quotation is not explained in the Soviet source, but the possibility suggests itself at once that reference is to subsidized losses, and accordingly that these totaled 10 billion rubles in 1937, or 2 billion rubles more than has been assumed in this study. The writer was disinclined to rely on this theory, however, chiefly because of doubts as to the scope of the figure on profits net of "losses." Particularly, in the context this may well refer not to the whole economy but to industry alone.

As has been explained previously, a major reform was instituted in the Soviet price system in April, 1936, declaredly with the major aim of establishing profitability as a general rule for Soviet economic organization. Contrary to the impression given by some Soviet discussions, the Soviet financial plan for 1937 itself makes clear that this reform by no means meant the complete elimination of subsidies, and our estimate of the realized figure assumes that the reform was even less successful than the plan forecast. But it may be in order to refer here, finally, to two other aspects concerning the success of the reform.

First, there was undoubtedly a very sizable reduction in subsidies. According to Gosplan, *Narodno-khoziaistvennyi plan na 1936 god* (National Economic Plan for 1936), p. 381, a total of 26.7 billion rubles was allocated to investments in working capital and subsidies in the financial plan for 1936. According to Smilga (p. 118), working capital actually increased by 9.9 billion rubles in that year, so the indications are that subsidies were to have been upwards of 15 billion rubles, or perhaps twice as much as we take the actual subsidies for 1937 to be. It is believed that the annual plan for 1936 was formulated without reference to this reform, so our calculations may be taken to measure the full effects of the reform, so far as subsidies are concerned.

Second, the concern about profitability was directed primarily to industry. At the same time, as will appear subsequently, a major part of the subsidies still granted in 1937 were to agriculture, particularly to the MTS and state farms; accordingly, for industry alone, the decline in subsidies was greater than our data on total subsidies show. The subsidies to the MTS, it should be noted, have a somewhat special character insofar as they arise from the Soviet practice of valuing the income in kind of the MTS at the low government procurement prices for agricultural produce. As was indicated previously, the MTS which previously had been on a *khozraschët* basis were attached directly to the government budget beginning in 1938. From that time on, the practice of referring to "subsidies" in connection with these organizations is somewhat out of order; and so I believe has been abandoned in Soviet budget practice.

On the various features of the Soviet budget classification referred to in the foregoing paragraphs see Narkomfin, *Klassifikatsiia raskhodov i dokhodov edinogo gosudarstvennogo biudzheta na 1936 g.* (Classification of the Expenditures and Revenues of the Unified Government Budget for 1936); *Prikaz NKF SSSR* (Order of the People's Commissariat of Finance, USSR), April 2, 1938, No. 163/121, *Finansovyi i khoziaistvennyi biulleten'* (Financial and Economic Bulletin), May 30, 1938, No. 15, pp. 3 ff.; G. A. Kozlov, ed., *Finansy i kredit SSSR* (Finance and Credit of the USSR), Ch. 14; N. N. Rovinskii, *Gosudarstvennyi biudzhet SSSR* (Government Budget of the USSR), 1944, ed., pp. 58–59, 66–67, 132–133, 250 ff.; 1949 ed., pp. 340–341; S. M. Kutyrev, *Analis balansa dokhodov i raskhodov khoziaistvennoi organizatsii* (Analysis of the Balance of Incomes and Outlays of the Economic Organization), p. 120.

*6. Depreciation.* This represents the amount of depreciation which the financial plan for 1937 estimates would be recorded in the books of Soviet enterprise in that year. See Bogolepov, p. 105.

*8. Transfer receipts.* See Table 1.

### B. OUTLAYS

*1. Communal services.* On the basis of per capita figures in TSUNKHU, *Sotsialisticheskoe stroitel'stvo Soiuza SSR* (Socialist Construction of the USSR), reprinted in *Planovoe khoziaistvo,* 1939, No. 8, p. 182, and assuming that the population of the USSR was 165 millions as of July 1, 1937, it is calculated that the total expenditures for education and health care in 1937 amounted respectively to 21.01 and 9.93 billion rubles. These calculated figures, it is believed, include capital investments, which according to the plan for 1937 were to amount to 1.45 billion rubles in the case of education and to 1.03 billion rubles in the case of health care (*Planovoe khoziaistvo,* 1937, No. 3, p. 223). The figures on education and health care cited in Table 3 are calculated as net of these respective capital outlays; also the figure on education in Table 3 is net of stipends which it is believed are included in the total of 21.01 billion rubles. The stipends have been estimated at 2.16 billion rubles (see Appendix A, item on stipends, p. 109). The figures obtained on education and health care are believed to cover not only expenditures out of the government budget but also expenditures by nongovernmental institutions (economic enterprises, trade unions, etc.).

The figure of 1.0 billion rubles for "Other communal services" is a quite arbitrary sum to allow for outlays on such items as physical culture, the administrative costs of the social insurance system (this is largely administered by the trade unions), etc., which presumably are not covered in items a and b.

2. *Government administration.* See Narkomfin, *Gosudarstvennyi biudzhet SSSR za vtoruiu piatiletku,* pp. 8–11.

3. *NKVD.* See *ibid.,* pp. 8–11.

4. *Defense.* The figure for this item, from Narkomfin, pp. 8–11, refers to the total outlays of both the Commissariat of Defense, so-called, and the Commissariat of the Naval Fleet (created in July 1937). It should be noted, however, that since the production of munitions at this time was largely, if not entirely, the responsibility of the Commissariat of Defense Industry, the budgets of the Commissariats of Defense and the Naval Fleet may not be comprehensive of all military outlays. Presumably the budgets *are* comprehensive in the case of finished munitions delivered to the Defense and Naval Fleet Commissariats; in the case of outlays for defense plant construction, however, insofar as this work is carried out by the Commissariat of Defense Industry, the outlays presumably would *not* be included in the budgets of the Defense Commissariat.

5. *Gross investment, etc.* This is calculated as a residual, i.e., 56.1 billion rubles is the difference between the sum of all incomes and the sum of all outlays other than gross investment. But reference may be made also to the following information on different types of investments actually carried out:

(i) Investments in fixed capital. According to *Second Session of the Supreme Soviet of the USSR, August 10–21, 1938; verbatim report,* p. 533, the total investments in fixed capital (*kapital'nye vlozheniia*) amounted to 29.5 billion rubles in 1937. According to *Tretiia Sessiia Verkhovnogo Soveta SSSR, 25–31 maia, 1939; stenograficheskii otchët* (Third Session of the Supreme Soviet of the USSR, May 25–31, 1939; verbatim report), p. 399, the total for this same item is given as 33.2 billion rubles. It is not known how to account for the difference between these two figures. Possibly the larger one is inclusive of and the smaller one exclusive of: (a) capital repairs not charged as current expense; (b) the "extra limit" investments which under Soviet planning practice are only partially controlled by the national economic plan, i.e., apparently the plan controls the total sums to be expended but not the concrete projects. On the basis of scattered information, it appears that these two items together might have amounted to several billion rubles. In the plan for 1937, the latter outlays were to total 2.6 billion rubles (Smilga, p. 118).

(ii) Investments in working capital. (a) According to data in Smilga, pp. 118–119; Grin'ko, pp. 38–39; Sokolov, p. 36; *Second Session of the Supreme Soviet,* p. 536, the working capital of economic organizations increased from 71 billion rubles on January 1, 1937 to 86 billion rubles on January 1, 1938, or by 15 billion rubles in the course of the year. These figures, it is believed, refer to the credit rather than debit side of the enterprises' ledger, in the sense that they represent bank loans, government grants, and other funds available for working capital needs, rather than working assets. Since the latter presumably includes some cash and unused bank balances, the increase in working capital as computed above probably exceeds the amount of investments in physical working assets. On the other hand, as has been explained (above, p. 117), reference may be only to state enterprises to the exclusion of co-operative enterprises, and there would be an understatement on this account.

(iii) Collective farm investments. The foregoing items, it is believed, do *not* include: (a) Collective farm investments out of income in kind, amounting, according to information presented previously, to 2.0 billion rubles; and (b) Collective farm money investments, calculated to amount to 2.8 billion rubles in 1937. According to information in I. V. Sautin, *Kolkhozy vo vtoroi Stalinskoi piatiletke,* p. 129, the money outlays by collective farms on capital construction, and the purchase of cattle and equipment amounted to about 11,500 rubles per collective farm in 1937. We take the total number of collective farms here to be 240,000.

(iv) Gold production. The gold production of the USSR is stated to have been between 4.35 and 5.40 million ounces in 1937 (A. Gerschenkron, *Economic Relations with the USSR,* p. 36). This would come to about one billion foreign trade rubles (5.3 rubles = $1); undoubtedly the actual ruble costs were several times this sum.

(v) Net foreign balance. The foreign balance on commodity account was favorable to the extent of 387 billion rubles in 1937 (again these are foreign trade rubles). No information is at hand regarding the balance on service account.

(vi) Commodity stockpiling. In the planned budget for 1937 some 1.7 billion rubles seem to have been appropriated for this purpose (Grin'ko, p. 72).

The foregoing items, in the aggregate, amount to some 56 billion rubles. In addition, reference should be made to the government outlays on such activities as prospecting, resettlement, etc. (see above, pp. 23–24), which would constitute investments as calculated here.

7. *Transfer outlays.* See Table 1.

# APPENDIX C. NOTES TO TABLES 5 AND 9

The elements in the calculation in Table 5 of the net national product by economic sector are shown in Appendix Table 1, columns 1 to 10. At the same time, the elements of the calculation in Table 9 of the national product by sector, allowing for taxes and subsidies, are set forth in Appendix Table 1, columns 1 to 11. Referring to this table, the sources and methods used in the calculations are as follows:

*Column 2. Earned income of households: Wages, TSUNKHU reports.* This tabulation is based on that in Gosplan, *Tretii piatiletnii plan,* pp. 228–29. The wage bill of the economic sector "Housing and communal economy" (*Zhilishchnoe i kommunal'noe khoziaistvo*) cited in this source, 1.74 billion rubles, has been distributed as follows among the sectors listed in Appendix Table 1: .44 billion rubles to "Industry and construction," to represent local gas and water works and similar industrial public services; .43 billion rubles to "Transportation and communications," to represent local tramways, etc.; and .87 to "Services, including government," to represent housing.

In addition to housing, the following sectors listed in the indicated source are included in Appendix Table I under the heading "Services, including government": Education; Cultural facilities (*Iskusstvo*); Health care; and Government and social institutions. The following sectors listed in the indicated source are included in Appendix Table 1 under the heading of "Other": Forestry (*Lesnoe khoziaistvo*); and Other. The nature and scope of the category "Forestry" referred to here is not clear. It is known that lumbering generally is included under the heading of "Industry."

In the indicated source, there is a gap of 3.37 billion rubles between the sum of the specified figures on wages by sectors, including "Other," and the specified wage bill for the whole economy. This gap is shown in Appendix Table 1 under the heading "Statistical discrepancy."

*Column 3. Earned income of households: Wages, other.* As is indicated in Appendix A, p. 108, "Other wages," not covered in the TSUNKHU reports, amount to .30 billion rubles in the case of agriculture and 21.13 billion rubles for all other sectors. It is assumed that of the latter sum, one sixth or 3.52 billion rubles, represents the wages of penal workers, all of whom are taken to be employed in "Industry and construction" (at a rate of pay equal to one half that of free workers this would allow for a penal labor force of about 2.5 millions); and 17.61 billion rubles represents wages omitted from the TSUNKHU data because of reporting deficiencies to which reference already has been made. It is assumed that as between sectors the amount of wages omitted from the TSUNKHU data on account of these reporting deficiencies is proportional to the amount of wages reported.

The assumptions made above concerning the wage bill, average wage, and total numbers of penal workers are arbitrary; for present purposes it hardly

## APPENDIX TABLE I

### CALCULATION OF NET NATIONAL PRODUCT BY ECONOMIC SECTOR, USSR 1937 [a]

*(All figures in billion rubles)*

| ITEM | EARNED INCOME OF HOUSEHOLDS | | | CONTRIBUTIONS TO SOCIAL INSURANCE | OTHER ALLOCATION TO SPECIAL FUNDS; MISC. INDIRECT TAXES AND FEES | NET INCOME BEFORE SUBSIDIZED LOSSES AND TAXES | TURNOVER TAXES | SUBSIDIZED LOSSES | NET NATIONAL PRODUCT [TOTAL OF COLS. (2) THROUGH (8), LESS (9)] | NET NATIONAL PRODUCT LESS TURNOVER TAXES, PLUS SUBSIDIES |
|---|---|---|---|---|---|---|---|---|---|---|
| | Wages, TSUNKHU reports | Wages, other | Incomes other than wages | | | | | | | |
| (1) | (2) | (3) | (4) | (5) | (6) | (7) | (8) | (9) | (10) | (11) |
| 1. Agriculture | 5.27 | .30 | 54.04 | .34 | .50 | 4.72 | | 4.50 | 60.67 | 65.17 |
| 2. Industry and construction | 37.08 | 12.00 | 9.15 | 3.59 | 2.74 | 11.49 | 47.66 | 2.50 | 121.21 | 76.05 |
| 3. Transportation, communications | 10.38 | 2.37 | .91 | .97 | .77 | 1.36 | | .50 | 16.26 | 16.76 |
| 4. Trade, including restaurants | 5.85 | 1.34 | | .38 | .43 | 2.40 | 27.56 | .50 | 37.46 | 10.40 |
| 5. Finance | .66 | .15 | | .04 | .05 | 1.00 | | | 1.90 | 1.90 |
| 6. Services, including government | 18.86 | 4.31 | 11.64 | 1.24 | 1.39 | .60 | | | 38.04 | 38.04 |
| 7. Other | .78 | .18 | | .05 | .06 | | | | 1.07 | 1.07 |
| 8. Statistical discrepancy | 3.37 | .77 | 4.28 | | .25 | | .69 | | 9.36 | 8.67 |
| 9. Total | 82.25 | 21.43 | 80.02 | 6.61 | 6.19 | 21.57 | 75.91 | 8.00 | 285.98 | 218.07 |

[a] Minor discrepancies between indicated totals and calculated sums of items are due to rounding.

seems possible even to survey the available information on this controversial subject. But, in order to suggest the range of conjecture, attention should be drawn to the following items:

(i) A Soviet representative at the United Nations informs us (*The New York Times*, October 17, 1948) that persons engaged in corrective labor in the USSR are paid up to 60 rubles a month. It is not indicated whether this includes the value of subsistence, but at current prices this could hardly be the case, even for the meager rations in question. The average wage for free labor in the USSR at the time of this release probably was in the neighborhood of 600–700 rubles a month.

(ii) D. J. Dallin and B. I. Nicolaevsky (*Forced Labor in Soviet Russia*, p. 86) estimate that the penal workers number about 7 to 12 million. In the course of their study, the authors examine a great deal of material on Soviet penal labor, particularly accounts of former penal workers who have made their way to the West; but they do not make clear just how this estimate is calculated.

(iii) In his *The Population of the Soviet Union*, p. 228, Frank Lorimer endeavors to calculate on the basis of various Soviet data the distribution of the Soviet population by principal occupation of the gainfully employed for the census date, January 17, 1939. The data in question include TSUNKHU data on wage earners and salaried workers, and various Soviet census data. On this basis, Lorimer is able to assign to different occupations all but 6.79 million persons, including gainfully employed together with their dependents. Since Lorimer does not take into account penal workers or other wage earners not covered in the TSUNKHU reports, the implication is that his unexplained residual represents the numbers of such workers, together with their dependents. At the same time, this indirect estimate of such workers is subject to appreciable error, owing to limitations which Lorimer makes clear are inherent in the basic data used.

(iv) Reference may be made finally to a secret Soviet source now available in this country, *Gosudarstvennyi plan razvitiia narodnogo khoziaistva SSSR na 1941 god* (State Plan for the Growth of the National Economy of the USSR for 1941). According to this document, which unquestionably is authentic, the NKVD was expected to produce in 1941, 1.97 billion rubles worth of industrial goods in 1926–27 prices, or 1.2 percent of the total output of all industry (pp. 9–10). The goal for the industrial labor force covered in TSUNKHU reports was 11.09 millions (p. 512), so one might suppose that the industrial employment by the NKVD would be of the order of only 150 thousand. But account has to be taken here of the fact that much of the NKVD industrial employment is in lumbering and extractive industries, where employment is relatively great in relation to output; also, the NKVD presumably is rather less concerned than other ministries with labor economies, so this relation must be even greater in the NKVD than in like industries. For a planned output of 3.54 billion rubles (p. 9), or less than twice that of the NKVD, the National Commissariat of Lumbering was to employ 1.30 millions (p. 521). At the same time, the NKVD is engaged in not only industry but also construction.

According to the plan, NKVD construction was to amount to 6.81 billion rubles, or 18.1 percent of the total volume of construction to be undertaken by all agencies (p. 484). It is believed that this represents, strictly speaking, the volume of construction undertaken for the NKVD either by this agency itself or by others. At the same time, the NKVD probably undertook some construction work for other agencies which would not be included in the indicated total. The goal for the entire construction labor force covered by TSUNKHU was 3.09 millions (p. 512).

*Column 4. Earned income of households: Incomes other than wages.*

(a) Agriculture. See Table 1 and the appended notes, according to which the money payments to collective farmers on a labor basis, salaries of collective farm executives and premiums, net money income of farm households from the sale of farm products, and net farm income in kind are found to total together 54.04 billion rubles.

(b) Industry and construction. This comprises: (i) The earnings of cooperative artisans, calculated to be 4.60 billion rubles. According to data in N. S. Margolin, *Balans denezhnykh dokhodov i raskhodov naseleniia,* p. 8, the total earnings of "cooperative artisans" were equal in 1934 to 3.44 percent and in 1938 to 3.60 percent of the wage and salary bill. On the basis of information on the wage and salary bill for 1934 and 1938 in Bergson, "A Problem in Soviet Statistics," p. 236, the earnings of "cooperative artisans" can be determined in absolute terms for these years. The cited figure for 1937 is obtained by interpolation.

(ii) The earnings of independent artisans in industrial pursuits, including domestic handicraft workers, taken to amount to 4.55 billion rubles. It was found in Appendix A, p. 108, that the total income of cooperative and independent artisans, together with domestics, haulers, etc., amounts to 13.70 billion rubles. After deducting the earnings of cooperative artisans, it is assumed here that the balance of 9.10 billion rubles is equally divided between the incomes of independent artisans engaged in industrial pursuits and the incomes of other independent artisans (barbers, etc.), domestics, haulers, etc. The latter incomes are included to the extent of one fifth under "Transportation, communications" and four fifths under "Services, including government."

(c) Transportation, communications. See the comment on "Industry and construction."

(d) Services, including government. Besides the incomes mentioned above in the comment on "Industry and construction," this includes the imputed net rent of owner-occupied houses, and military pay and subsistence, as estimated in Appendix A.

(e) Statistical discrepancy. See Appendix A, p. 109.

*Column 5. Contributions to social insurance.* In a decree of the Council of Commissars of March 23, 1937, is set forth the schedule of social insurance contribution rates effective as of January, 1937, for workers employed in the

121 different industrial unions of the USSR. See *Sobranie zakonov i rasporia-zhenii SSSR* (Collected Laws and Decrees of the USSR), 1937, Part I, No. 22, Section 88. Assuming provisionally that unweighted average rates computed for the unions in the different economic sectors listed in Appendix Table 1 apply to the TSUNKHU wage bill, the following calculation is made:

|  | CALCULATED AVERAGE RATE OF CONTRIBUTION, AS PERCENT OF WAGES | CALCULATED CONTRIBUTIONS, BILLIONS OF RUBLES |
|---|---|---|
| Agriculture | 4.8 | .25 |
| Industry and construction | 7.1 | 2.63 |
| Transportation and communications | 6.8 | .71 |
| Trade, including restaurants | 4.8 | .28 |
| Finance | 5.0 | .03 |
| Services, including government | 4.8 | .91 |
| Other | 5.0 [a] | .04 |

[a] An arbitrary rate.

As computed here, the total of the social insurance contributions from all sectors, is 4.85 billion rubles. Actually, however, these contributions totaled 6.61 billion rubles in 1937 (Appendix B). In Appendix Table 1, the contributions of the different sectors as computed above have been adjusted upwards in proportion to the difference between the computed and the actual totals.

*Column 6. Other allocations to special funds; miscellaneous indirect taxes and fees.* According to Appendix B, this item totals 6.19 billion rubles. Of this total, it is assumed arbitrarily that .5 billion rubles represents the allocations and payments made by agriculture. The balance is distributed as between other sectors in proportion to the amount of wages reported by TSUNKHU.

*Column 7. Net income before taxes.* According to Appendix B, the net income of collective farms before taxes amounts to 4.57 billion rubles (including retained income in kind and money income and income taxes) and the net profit of all other enterprises before taxes amounts to 17.00 billion rubles.

From data in Smilga, "Finansy sotsialisticheskogo gosudarstva" (Finances of the Socialist Government), *Problemy ekonomiki,* 1937, No. 2, p. 112, we may compute the following tabulation of the *planned* profits by sectors for the year 1937:

| SECTOR | | PROFITS, BILLION RUBLES |
|---|---|---|
| 1. Agricultural enterprises other than collective farms | | .19 |
| 2. Industry and construction | | |
| a. Other than cooperatives and local utilities | 12.60 | |
| b. Cooperatives | 1.98 | |
| c. Local utilities | .39 | |
| d. Total | | 14.97 |

| SECTOR | | PROFITS, BILLION RUBLES |
|---|---|---|
| 3. Transportation and communications | | |
| a. Other than local | 1.39 | |
| b. Local | .38 | |
| c. Total | | 1.77 |
| 4. Trade, including restaurants | | |
| a. State | 2.00 | |
| b. Cooperative | 1.13 | |
| c. Total | | 3.13 |
| 5. Services, including government | | .78 |
| 6. Total | | 20.84 |

In compiling the foregoing tabulation from the data in Smilga, it is arbitrarily assumed that the item "Communal services and housing" listed in this source comprises the profits of local utilities, in the amount of one fourth the total; of local transport facilities, in the amount of one fourth the total; and of housing, in the amount of one half the total. Also it is assumed that profits in construction, not listed by Smilga, are zero.

Smilga does not include any figure for the profits of financial institutions either. From other information, it is clear that these institutions did earn substantial profits in 1937, and it is assumed here that these profits in fact totaled 1.0 billion rubles. (According to Narkomfin, *Gosudarstvennyi biudzhet SSSR za vtoruiu piatiletku*, p. 8, financial institutions in 1937 paid .612 billion rubles out of profits into the government budget.)

Accordingly, of the total of 17.00 billion profits *actually* earned by enterprises other than the collective farms during 1937 (see above, Appendix B), it is assumed that 1.00 billion accrued to financial institutions and 16.00 to other sectors listed by Smilga. As is noted, the profits of the latter sectors were *planned* to amount to 20.84 billion rubles. It is assumed that for each of the sectors listed by Smilga the realized profits bear the same relation to planned profits as they do for all sectors together.

*Column 8. Turnover taxes.* According to Grin'ko, p. 69, the expected yield of the turnover tax in 1937 was, for enterprises subject to the Commissariats of Heavy, Light and Food Industry, 46.82 billion rubles; for enterprises under the Council of Commissars' Committee for the Procurement of Agricultural Products, 24.11 billion rubles; for enterprises in state trade, 2.61 billion rubles; and in the case of cooperatives, 1.68 billion rubles. It is assumed that these forecasts actually were realized and that, in the case of the cooperatives, the total yield from the tax is divided equally between those in industry and those in trade (in which category the taxes paid by procurement agencies also are classified). The total revenue from the turnover tax from all sources actually turned out to be 75.91 billion rubles (Appendix B) or .69 billions greater than the sum of the forecasted yields from the specified sectors.

*Column 9. Subsidized losses.* Elsewhere the total of subsidized losses has been estimated at 8.0 billion rubles, comprising 7.5 billion rubles covered in the consolidated government budget heading referring to subsidies generally, probably outlays "on mastering production," and .5 billion rubles representing certain payments to procurement agencies presumed to be classified in the budget under the heading of "operational outlays" (Appendix B, pp. 114 ff.). At the same time, these payments are included with various others under the more general heading of outlays on the "National Economy." The other payments included are budget investments in fixed capital, other than "extra-limit outlays"; budget appropriations for "extra-limit outlays"; budget investments in "own" working capital; budget investments in commodity reserves; and various other outlays, chiefly on industrial resettlement and special measures of a like sort. These last constitute the main part of "operational outlays."

In the first five columns of Appendix Table 2 are shown some data that have been compiled on the total outlays on the "National Economy" by economic sector and on a number of its elements. Thus, in column 1 are data on the total outlays. These are taken directly from Narkomfin, p. 9. In column 2 are data on the budget investments in fixed capital, from *ibid.*, p. 76. In column 3 are some estimates of budget appropriations for "extra-limit" outlays, based on data in Smilga, p. 118. (Smilga's data refer to the total of "extra-limit outlays" by economic sector, including both outlays financed out of the budget and outlays financed from other sources. However, a breakdown of this total by source is given for all sectors together. I have assumed that the breakdown for individual sectors is in the same proportion as for all sectors together. The breakdown within the transportation and communications sector as between railway and other sectors is arbitrary.) In column 4 are some data on the all-union budget investments in "own" working capital, from Smilga, p. 119. Finally, in column 5 there is a figure on budget investments in commodity reserves, from Grin'ko, p. 72.

As is indicated in the table, while the data on total outlays and those on fixed capital represent realized figures, the remaining data represent planned figures. Accordingly, the balance of realized outlays not yet accounted for represents the total of differences between planned and realized appropriations for "extra-limit" outlays, working capital, and reserves; together with "operational expenditures" on special measures and other miscellaneous expenditures; and subsidies. (For present purposes I am assuming that the data on investments in fixed capital are exclusive of "extra-limit" outlays; this is not entirely clear in the Soviet source, so there may be some double-counting in our calculations at this point.)

For purposes of calculating subsidies, let me now explain, I have allocated this balance as between elements other than subsidies, on the one hand, and subsidies, on the other, so as to take account of various considerations, chiefly:

(i) Prior to the price reform of April, 1936, the major recipient of subsidies by all accounts was industry. At the same time, the price reform was directed chiefly at this sector. But there is no question that an appreciable amount of subsidies still was paid to industry after the reform. On the price reform of April,

## APPENDIX TABLE 2

## CONSOLIDATED GOVERNMENT BUDGET OUTLAYS ON "NATIONAL ECONOMY," USSR 1937

*(Billions of rubles)*

| | REALIZED TOTAL OUTLAYS | REALIZED INVESTMENTS IN FIXED CAPITAL | PLANNED "EXTRA-LIMIT" OUTLAYS | PLANNED INVESTMENTS IN WORKING CAPITAL, ALL-UNION BUDGET ONLY | PLANNED RESERVES | ASSUMED BALANCE, OTHER THAN SUBSIDIES | ASSUMED SUBSIDIES |
|---|---|---|---|---|---|---|---|
| | (1) | (2) | (3) | (4) | (5) | (6) | (7) |
| 1. Agriculture | 9.51 | 1.55 | .28 | 1.33 | | 1.85 | 4.50 |
| 2. Industry | 16.74 | 8.90 | { .18 / .11 | 3.30 | | 1.75 | 2.50 |
| 3. Transportation and communications | | | | | | | |
| a. Railways | 4.33 | 3.54 | .04 | .12 | | .63 | 0 |
| b. Other transport, communications | 3.77 | 2.08 | .03 | .41 | | .75 | .50 |
| c. Total | 8.10 | 5.62 | .07 | .53 | | 1.38 | .50 |
| 4. Trade | 3.16 | .20 | { .08 / .11 | .20 | | 2.07 | .50 |
| 5. Communal and housing construction | 2.76 | 1.89 | .54 | | | .33 | 0 |
| 6. Other | 3.14 | .10 | .01 | .17 | 1.69 | 1.17 | 0 |
| 7. Total | 43.41 | 18.26 | 1.38 | 5.53 | 1.69 | 8.55 | 8.00 |

1936, see *Izvestiia,* April 11, 1936. On the postreform subsidies to industry, see Gosplan, *Tretii piatiletnii plan,* p. 114; L. Vilenskii, "Finansovye voprosy promyshlennosti" (Financial Problems of Industry), *Planovoe khoziaistvo,* 1938, No. 10, pp. 57–60; Chernomordik, ed., *Narodnyi dokhod SSSR* pp. 82–84;

(ii) The government practice of valuing the income in kind of MTS at the low government procurement prices necessarily means very sizable subsidies to this sector after as well as before the price reform. Apparently some subsidies had to be paid, too, to the state farms. See C. Glezin, compiler, *Biudzhetnaia sistema Soiuza SSR* (Budget System of the USSR), pp. 244 ff.; A. K. Suchkov, *Dokhody gosudarstvennogo biudzheta SSSR,* Ch. VII; N. N. Rovinskii, *Gosudarstvennyi biudzhet SSSR,* 1944 ed., pp. 157 ff.;

(iii) According to information in Soviet sources, water transport continued to suffer notable losses after the price reform. Except for railroads, there may have been losses, too, in other branches of transport. It is believed that in the case of railways, losses as of 1937 were covered out of profits cleared through the Commissariat of Railways rather than from subsidies from the government budget. This is discussed in the separate memorandum on the Soviet practice of accounting for subsidized losses, referred to above; see p. 113. On losses in transportation, see Bogolepov, p. 109; Smilga, p. 112; Gosplan, *Tretii piatiletnii plan,* pp. 114–115.

(iv) The sum of .5 billion rubles taken here to represent subsidies to procurement agencies falls under "Trade" in our table. No reference has been found in Soviet literature to other subsidies to trade.

(v) No reference has been found in Soviet literature to subsidies to other sectors than the foregoing.

(vi) "Operational outlays" on industrial resettlement and similar measures apparently are concentrated to a considerable extent in industry and agriculture.

(vii) In view of the large measure of republican and local responsibility for trade, there probably were sizable investments in working capital in this sector financed out of republican and local budgets. Such funds are not covered by the data in column (iv) of our table.

Reference should be made finally to the puzzling passage from Gosplan, *Tretii piatiletnii plan,* p. 114, cited above, p. 118. According to one interpretation of this passage, the profits of Soviet industry after subsidized losses were 7 billion rubles in 1937. At the same time, the profits of transportation and communications seem to have been .1 billion rubles. (In the context, *"nakopleniia"* very likely stands for profits.) Taken together with our estimates of gross profits before subsidized losses in Appendix Table 1, the data compiled here on subsidies indicate net profits amounting to some 9 billion rubles for industry and to .86 billion rubles for transportation and communications. Each of these sums includes about one quarter of a billion rubles of profits of local utilities. Probably the profits of such organizations are not taken into account in the Gosplan figures. But even so the indication is that as a result of either an understatement of subsidized losses or an overstatement of gross profits before such losses, our estimates of net profits may be too high, by nearly 2 billion rubles in the case of industry and by about .5 billions in the case of transportation and communications.

# APPENDIX D. NOTES TO TABLE 6

A. REVENUES

*1. Direct taxes.* Narkomfin, *Gosudarstvennyi biudzhet SSSR za vtoruiu piatiletku,* pp. 8–11.

*2. Net borrowing.* Computed from data in Baykov, *The Development of the Soviet Economic System,* p. 380. Baykov's tabulations in turn are taken from D'iachenko, *Finansy i kredit SSSR,* pp. 264–68.

*3. Statistical discrepancy between budget data on current loan transactions and other data on outstanding debt.* According to the budget for 1937 (Narkomfin, pp. 8–11), the total revenues on loan account amounted to 5.87 billion rubles and the total outlays, presumably including both amortization and interest, to 3.46 billion rubles. This means that the net revenues from new loans, over and above amortization and interest, amounted to 2.41 billion rubles. According to the data in Baykov just cited, the total outstanding debt increased on balance by 2.21 billion rubles during 1937; but, if interest at 4 percent is allowed on the outstanding debt, the net revenue from new loans, over and above interest and amortization, comes to only 1.10 billion rubles.

To explain the discrepancy, a variety of possibilities suggest themselves, among them that Baykov's data on the outstanding debt may not be comprehensive and in particular may not include, say, short term borrowing from the banking system. Unfortunately there are no data at hand to check this theory.

*4. Revenues of social insurance budget.* Narkomfin, pp. 8–11, 77; Plotnikov, *Biudzhet Sovetskogo gosudarstva,* p. 75. According to Soviet practice, only a part of the revenue from the social insurance tax is assigned to social insurance agencies for expenditure. The balance is retained by the government, as a part of the general governmental revenues. Until 1938, the Soviet budget practice was to include only the latter part among the revenues in the government budget; beginning in that year, however, the entire revenue from the payroll tax was listed as a revenue item in the government budget. The Narkomfin publication cited here follows the pre-1938 practice; in Plotnikov, however, the budget for 1937 has been revised in accord with the post-1938 practice, and I have adopted this same procedure here.

*5. Indirect taxes, etc.* See Appendix B, p. 114.

*6. Other.* This is the residual revenue of 3.25 billion rubles in the budget of 1937 (Narkomfin, pp. 8–11) after account is taken of all items included under other headings—including the miscellaneous indirect taxes, etc.—and allowance is made for the fact that our data on loan transactions are on a net basis while those in the budget are on a gross basis. As has been indicated, I assume that .91 billion rubles of the residual revenue represents indirect taxes and fees, e.g., fines paid by organizations, auto inspection fees, etc. For purposes of this study, the balance of 2.34 billion rubles is treated as comprising transactions on capital account, e.g., refunds of budget outlays of past years. Actually, this procedure is something of an oversimplification. For one thing, some of the revenues, e.g., fines on individuals, represent direct taxes and charges, and should have been allowed for in Table 1. For another, revenues from loans issued by local government units apparently are included here rather than in the loan account so-called in the government budget. Whether and to what extent such revenues have been taken into account in Table 1 is not known. The residual item probably includes mintage revenues; however, I believe these are properly omitted, along with capital transactions, both from direct and indirect taxes and fees. On the contents of the residual revenue item, see Narkomfin, p. 5; also *Prikaz NKF SSSR,* April 2, 1938, No. 163/121, *Finansovyi i khoziaistvennyi biulleten',* May 30, 1938, No. 15, pp. 3 ff.

B. EXPENDITURES

*1. Interest charges on debt.* These data are computed so as to allow 4.0 percent on the outstanding debt as shown in Baykov, *The Development of the Soviet Economic System,* p. 380.

*2. Pensions and allowances, including those paid by the social insurance system.* See Appendix A, item on Pensions and Allowances, p. 109.

*3. Communal services.* Narkomfin, *Gosudarstvennyi biudzhet SSSR za vtoruiu piatiletku,* pp. 8–11.

*4. Government administration.* Narkomfin, pp. 8–11.

*5. NKVD.* Narkomfin, pp. 8–11.

*6. Defense.* Narkomfin, pp. 8–11.

*7. Financing the national economy.* Narkomfin, pp. 8–11.

*8. Other.* This is obtained as the difference between: (a) the sum of (i) government expenditures on "social-cultural measures" *other than education and health care,* as recorded in the budget, 2.31 billion rubles (Plotnikov, *Biudzhet Sovetskogo gosudarstva,* p. 79); (ii) expenditures of the social insurance system, as recorded in the budget, 5.27 billion rubles (Narkomfin, p. 77; Plotnikov, p. 79); and (iii) the item "Other expenditures," as recorded in the budget, 3.52 billion rubles (Narkomfin, pp. 8–11; Plotnikov, p. 79)—which together amount to 11.10 billion rubles; and (b) the total expenditures on pensions and allowances, amounting to 6.10 billion rubles.

As thus computed, the item "Other expenditures" comprises:

(a) All outlays of the social insurance system other than pensions and allowances; in other words, all outlays of the social insurance system for administration, health care, etc.

(b) Miscellaneous budget outlays for "social-cultural measures" other than for pensions and allowances, e.g., for physical culture, possibly for parks, etc.

(c) Miscellaneous expenditures. In this study these are treated as the counterpart of transactions on capital account included under the heading of "Other revenues." This treatment apparently is in order for some of the items included, e.g., transactions presumably on capital account with the State Bank and with the long-term investment banks; but again as with the procedure regarding "Other Revenues" it represents something of an oversimplification. Thus, outlays on local loan accounts are included here rather than in the loan-account so-called in the budget, and it is not known whether such outlays have been taken into account in Table 1. Also the residual item includes expenses of currency production; possibly such expenses should be included under the heading of government administration. On the contents of the residual item, see the sources cited above on "Other revenues."

9. *Indicated budget surplus*. The excess of revenues over expenditures, as reported in the state budget (Narkomfin, pp. 8–11; Plotnikov, pp. 75, 79).

# APPENDIX E. NOTES TO TABLE 8

The calculation of the data in terms of adjusted prices in Table 8 may be explained by reference to Appendix Table 3. The contents of this latter table are discussed by columns:

*Column 2. Value at established prices.* All figures, except those in parentheses, are taken from Tables 1 and 2. The figures in parentheses are obtained as follows:

6. *Communal services.* According to data in Gosplan, *Tretii piatiletnii plan,* pp. 228–29, the total wages paid out by educational, health, and cultural institutions in 1937 amounted to 11.13 billion rubles. If account is taken of the wages paid out by trade unions and other social organizations, it is believed the total would come to about 12.00 billion rubles. According to data for March, 1936, in TSUNKHU, *Chislennost' i zarabotnaia plata rabochikh i sluzhashchikh v SSSR* (Number and Wages of Wage Earners and Salaried Workers in the USSR), pp. 8–12, the total wages paid out by "social organizations" amounted to 10.7 percent of the total wages paid out by government institutions, including organs of economic administration and the judiciary, *and* the social organizations taken together. According to data in Gosplan, pp. 228–29, the total wages paid out by government institutions and social organizations in 1937 amounted to 6.86 billion rubles. The total of 12.00 billion rubles has been increased by 1.70 billion rubles, in other words, to one half the total outlays for communal services as a whole, in order to allow for possible incompleteness in the Gosplan (in reality, TSUNKHU) wage data in the various spheres referred to (see Appendix A, pp. 107 ff.). The remaining half of the outlays on communal services, it is assumed, represents expenditures on commodities.

7. *Government administration.* According to Gosplan, pp. 228–29, the total wages paid out by government institutions and social organizations in 1937 amounted to 6.86 billion rubles. For government institutions alone, it is believed that the total would come to about 6.00 billion rubles. (See item on communal services, above.) It is assumed that this total includes the wage bill not only of government institutions generally but also of the NKVD, and that exclusive of the NKVD the wage bill of government institutions would amount to about 60 percent of the indicated total, or 3.60 billion rubles. This assumes that the magnitudes of the wage bill in government institutions generally and in the NKVD are proportional to the total outlays of government institutions generally and of the NKVD. Government outlays on commodities are calculated as a residual.

8. *NKVD.* See above, note to item 7, "Government administration."

9. *Defense.* Military pay and army subsistence already have been taken to amount respectively to 1.50 and 2.50 billion rubles (Appendix A). The value of munitions is computed as a residual.

10. *Gross investment.* On the basis of the data assembled on this item in Ap-

pendix B, it is assumed here that aside from collective farm investment out of income in kind 30 percent of our gross investment, or 16.2 billion rubles, represents additions to commodity inventories and stockpiles, exclusive of gold, and that the remaining 70 percent, or 37.9 billion rubles, represents investments in fixed capital, current gold output for monetary purposes, and the net foreign balance.

*Columns 3 and 4. Turnover tax.* These data are intended to represent the amount of the turnover tax levied on the specified categories of goods and services, either directly or indirectly, insofar as taxed materials and fuel are used in their production. While for the most part the data were obtained by rule-of-thumb procedures, it is believed that they represent reasonably well the magnitudes involved, so that the final results of our adjustment are not likely to be very appreciably distorted by errors at this point.

The main steps in the calculation were as follows:

(i) It is known that no turnover tax is levied on collective farm market sales and, of course, there would have been no payments on trade union and other dues, farm income in kind, or labor services employed in the provision of communal services, in government administration, the NKVD, and the army. The adjustment in army pay is explained below. For the moment, for the purpose of computing the average effective tax rate referred to below, the taxes borne by army subsistence are taken into account only when this item appears in the defense budget and not when it appears as an element in the consumption of households.

(ii) The tax-free share of the gross national product thus amounts to 75.3 billion rubles. Since a sum of 75.9 billion rubles was collected, directly or indirectly, on the remaining part of the gross national product, amounting to 216.5 billion rubles, it is calculated that the average effective tax rate was 35.1 percent.

(iii) It is assumed that the following items are subject to this over-all average effective rate: all commodities used in the provision of communal services, government administration, NKVD, and army subsistence, and commodities included in inventories and stockpiles.

(iv) An average effective rate of 1.0 percent is assumed to apply to the item "Housing, services," to allow for the taxes levied on electric power, fuel, movie tickets, and possibly other items. According to Narkomfin, *Alfavitnyi perechen' promtovarov po stavkam naloga s oborota i biudzhetnykh natsenok* (Alphabetic List of Industrial Commodities by Rates of the Turnover Tax and Budget), electric power sold to both industrial and nonindustrial consumers was subject to a tax of 3.0 percent outside of those supplied by the Moscow and Leningrad regional stations, and to a tax of 23 percent in the latter areas; coal was subject to a tax of .5 percent; and fuel oils to a tax of 60.5 percent. These rates are percentages of the wholesale price gross of the tax and, except for the rates for electric power which were put in force in the spring of 1937, were effective throughout the year.

(v) An average effective rate of 7.5 percent is assumed to apply to munitions

## APPENDIX TABLE 3

### CALCULATION OF GROSS NATIONAL PRODUCT BY USE IN TERMS OF ADJUSTED RUBLE PRICES, 1937 [a]

| ITEM (1) | VALUE AT ESTABLISHED PRICES, BILLION RUBLES (2) | TURNOVER TAX Assumed effective rate Percent (3) | TURNOVER TAX Total billion rubles (4) | VALUE LESS TURNOVER TAX (5) | SUBSIDIES (6) | VALUE LESS TURNOVER TAX PLUS SUBSIDIES (7) | ADJUSTMENT FOR FARM PRICES (8) | VALUE AT ADJUSTED PRICES (9) |
|---|---|---|---|---|---|---|---|---|
| 1. Retail sales to households | | | | | | | | |
| a. Government and cooperative shops and restaurants | 111.5 | 53.8 | 60.0 | 51.5 | 4.1 | 55.6 | 5.3 | 60.9 |
| b. Collective farm markets | 16.0 | 0 | 0 | 16.0 | 0 | 16.0 | (—)7.3 | 8.7 |
| c. Total | 127.5 | 47.1 | 60.0 | 67.5 | 4.1 | 71.6 | (—)2.0 | 69.6 |
| 2. Housing; services | 19.9 | 1.0 | .2 | 19.7 | 0 | 19.7 | 0 | 19.7 |
| 3. Trade union and other dues | 1.1 | 0 | 0 | 1.1 | 0 | 1.1 | 0 | 1.1 |
| 4. Consumption of income in kind | | | | | | | | |
| a. Farm income | 32.5 | 0 | 0 | 32.5 | 0 | 32.5 | 0 | 32.5 |
| b. Army subsistence | 2.5 | 35.1 | .9 | 1.6 | .1 | 1.7 | .1 | 1.8 |
| c. Total | 35.0 | 2.6 | .9 | 34.1 | .1 | 34.2 | .1 | 34.3 |
| 5. Total outlays of households on goods and services | 183.5 | 33.3 | 61.1 | 122.4 | 4.2 | 126.6 | (—)1.9 | 124.7 |
| 6. Communal services | | | | | | | | |
| a. Commodities | (13.7) | 35.1 | 4.8 | 8.9 | .5 | 9.4 | .7 | 10.1 |
| b. Services | (13.7) | 0 | 0 | 13.7 | 0 | 13.7 | 0 | 13.7 |
| c. Total | 27.4 | 17.5 | 4.8 | 22.6 | .5 | 23.1 | .7 | 23.8 |

| | | | | | | | | |
|---|---|---|---|---|---|---|---|---|
| **7. Government administration** | | | | | | | | |
| a. Commodities | (.8) | 35.1 | .3 | .5 | 0 | .5 | 0 | .5 |
| b. Services | (3.6) | 0 | 0 | 3.6 | 0 | 3.6 | 0 | 3.6 |
| c. Total | 4.4 | 6.8 | .3 | 4.1 | 0 | 4.1 | 0 | 4.1 |
| **8. NKVD** | | | | | | | | |
| a. Commodities | (.6) | 35.1 | .2 | .4 | 0 | .4 | 0 | .4 |
| b. Services | (2.4) | 0 | 0 | 2.4 | 0 | 2.4 | 0 | 2.4 |
| c. Total | 3.0 | 6.7 | .2 | 2.8 | 0 | 2.8 | 0 | 2.8 |
| **9. Defense** | | | | | | | | |
| a. Commodities for army subsistence | 2.5 | 35.1 | .9 | 1.6 | .1 | 1.7 | .1 | 1.8 |
| b. Munitions | (13.5) | 7.5 | 1.0 | 12.5 | .7 | 13.2 | .1 | 13.3 |
| c. Services | 1.5 | 0 | .9 | 2.4 | (—).1 | 2.3 | (—).1 | 2.2 |
| d. Total | 17.5 | 5.7 | 1.0 | 16.5 | .7 | 17.2 | .1 | 17.3 |
| **10. Gross investment** | | | | | | | | |
| a. Collective farm investment of income in kind | 2.0 | 0 | 0 | 2.0 | 0 | 2.0 | 0 | 2.0 |
| b. Commodity inventories and stockpiles | (16.2) | 35.1 | 5.7 | 10.5 | .7 | 11.2 | .5 | 11.7 |
| c. Other | (37.9) | 7.5 | 2.8 | 35.1 | 1.9 | 37.0 | .5 | 37.5 |
| d. Total | 56.1 | 15.2 | 8.5 | 47.6 | 2.6 | 50.2 | 1.0 | 51.2 |
| **11. Gross national product** | 291.8 | 26.0 | 75.9 | 215.9 | 8.0 | 223.9 | 0 | 223.9 |

a Minor discrepancies between indicated totals and calculated sums of items are due to rounding.

and fixed capital and other investment. According to the schedule of the Commissariat of Finance, just cited, the rates on heavy industrial goods generally are low, running from 0.5 percent in the case of coal and steel to 1.0 percent in the case of industrial machinery. The comparatively high rate of 7.5 percent is used partly to allow for the compounding that results when one taxed item is used to produce another, and partly to allow for the high taxes on petroleum products generally, including rates of 50.0 percent for benzol, 60.5 percent for fuel oils and 88.3 percent in the case of kerosene. The oil industry alone paid 8 percent of the total of all turnover taxes in 1939 (Suchkov, *Dokhody gosudarstvennogo biudzheta SSSR*, p. 16). If the same percentage applied in 1937, the yield would have been about 6 billion rubles. Of course a part of this sum is ultimately to be charged against retail sales, housing, etc., rather than against investment.

(vi) Using all the foregoing rates, it is possible to account for 15.9 billion rubles of the tax. The balance, 60.0 billion rubles, is assumed to fall directly or indirectly on the retail sales of consumers' goods to households by state and cooperative shops and restaurants.

In connection with the foregoing calculations, attention should be called here to some data on the expected yield of the turnover tax for 1937 included in the budget for that year by G. F. Grin'ko, *Finansovaia programma Soiuza SSR na 1937 god* (Financial Program of the USSR for 1937), p. 69. According to Grin'ko the tax was to be collected in the following amounts from the different administrative sectors:

|                                                        | BILLION RUBLES |
| ------------------------------------------------------ | -------------- |
| Commissariat of Heavy Industry                         | 8.9            |
| Commissariat of Light Industry                         | 11.4           |
| Commissariat of Food Industry                          | 20.4           |
| Chief Administration for the Alcohol Industry          | 6.2            |
| Committee for Procurements of Agricultural Products     | 24.1           |
| State Trade                                            | 2.6            |
| Cooperatives                                           | 1.7            |
| Other                                                  | 1.5            |
| Total                                                  | 76.8           |

Grin'ko does not explain the residual item of 1.5 billion rubles; possibly it refers to the Commissariat of Defense Industry.

In comparing Grin'ko's data with those presented here, it should be noted that probably an appreciable part of the taxes paid by the Commissariat of Heavy Industry was levied on consumers' goods produced in this industry (radios, houseware, etc.). In our tabulation, such taxes would be included among those falling on retail trade. The bulk of the taxes paid by the Committee on Procurements would be disposed of similarly in our tabulation, though presumably a part would be allocated to other use categories than retail trade.

In the case of army subsistence, the turnover tax is deducted under the heading of both household consumption and defense. An upward adjustment is made, however, in defense pay exactly corresponding to the reduction in army subsistence, so the nominal value of army services is constant. This will be explained later.

*Column 6. Subsidies.* Subsidized losses already have been estimated at 8 billion rubles for the economy as a whole. This total includes subsidized losses amounting to 4.5 billion rubles in agriculture; to 2.50 billion rubles in industry; and to 1.0 billion rubles in trade (Appendices B and C, pp. 114 ff. and 128 ff.). For purposes of allocating these amounts by use categories, reference is made to the following classification of these categories:

|  | VALUE IN 1937, BILLION RUBLES |
|---|---|
| Group I | |
| Defense, munitions | 13.5 |
| One half the investment in inventories and stockpiles | 8.1 |
| Fixed capital and other investment | 37.9 |
| Total | 59.5 |
| | |
| Group II | |
| State and cooperative retail trade | 111.5 |
| Commodities used in: | |
|     Communal services | 13.7 |
|     Government administration | .8 |
|     NKVD | .6 |
|     Military subsistence | 2.5 |
| One half the investment in inventories and stockpiles | 8.1 |
| Total | 137.2 |
| | |
| Group III | |
| Collective farm market trade | 16.0 |
| Farm income in kind | 32.5 |
| Collective farm investment in kind | 2.0 |
| Services used in: | |
|     Communal services | 13.7 |
|     Government administration | 3.6 |
|     NKVD | 2.4 |
|     Defense, including value of army subsistence | 4.0 |
| Housing and services | 19.9 |
| Trade union and other dues | 1.1 |
| Total | 95.2 |

In view of the nature of the Soviet policy on subsidies and the kinds of use categories involved, it seems safe to proceed here on the assumption that subsidies were granted directly or indirectly only on use categories in Groups I and

II and not at all on those included in Group III. Note again, regarding "Military subsistence," that in Group III this is taken as a part of the value of military services, and while the subsistence itself is affected by the subsidies the total value of these services is not. Paralleling the procedure for the turnover tax, a downward adjustment in pay is made of the same amount as the upward adjustment in subsistence.

The allocation of subsidies, then, proceeds as follows:

(i) I assume that of the total subsidies on agriculture, 4.5 billion rubles, 10 percent or .45 billions represents the subsidies on agricultural materials used to produce munitions and investment goods, and accordingly falls on Group I; and 90 percent or 4.05 billion rubles represents the subsidies on agricultural materials used to produce consumers' goods and other commodities, and accordingly falls on Group II. Within each group the subsidies are distributed by use category in proportion to the indicated values.

(ii) In the case of the subsidies on industry, amounting to 2.5 billion rubles, it is believed that these consisted almost exclusively of subsidies on basic industrial goods, e.g., coal. Accordingly, I assume that 90 percent of these subsidies, or 2.25 billion rubles, represent subsidies falling directly or indirectly on Group I; and 10 percent, or .25 billion rubles, represent subsidies falling directly or indirectly on Group II. Within each group the subsidies are distributed by use categories in proportion to the indicated value.

(iii) It is believed that the subsidies on trade, amounting to .5 billion rubles, are mainly if not exclusively subsidies on agricultural procurements. These are distributed by use categories in the same way as the subsidies on agriculture.

(iv) Subsidies to transport are taken to be .5 billion rubles. I allocate one half of this to Group I and the other half to Group II.

*Column 8. Adjustment for farm prices.* The adjustment is carried out under the following assumptions:

(i) Collective farm market prices, while not subject to the turnover tax or subsidies, decline proportionately to retail prices when the turnover taxes and subsidies are eliminated.

(ii) Procurement prices for farm produce are increased over and above the amounts of subsidies in order to compensate farmers for the cut in collective farm market prices. (To simplify the calculation, it is assumed that the money costs of seed, etc., used on the farm are unchanged throughout.)

(iii) The value of military services, including subsistence as presently recorded, is constant.

Regarding the last assumption, it should be noted that the value of military services as recorded in our initial calculation of the national product is taken as a datum throughout. The value per soldier comes to about 2,300 rubles per annum, which may be compared with the average wage of ordinary workers in 1937, 2,820 rubles per annum.

The calculation proceeds as follows: (i) In revaluing the national product for subsidies in column 6 (pp. 139–40) we have in effect calculated among other things the incidence on different use categories of a rise in procurement prices which would increase the total value of agricultural income by 5.0 billion rubles. The increase falls on different use categories in the following proportions:

|  | PERCENT |
|---|---|
| State and cooperative retail trade | 72.7 |
| Army subsistence | 1.8 |
| Commodities used in: |  |
| Communal services | 9.1 |
| Government administration | 0 |
| NKVD | 0 |
| Defense, munitions | 1.8 |
| Inventories and stockpiles | 7.3 |
| Fixed capital; other investments | 7.3 |
| Total | 100.0 |

For all other use categories, there is no increase.

(ii) From this it follows that the further increase in procurement prices needed to compensate for the cut in collective farm market prices falls on the different use categories in the same proportions.

(iii) Let $k$ be the percentage decline in retail prices in the state and cooperative shops after all adjustments. As has been mentioned, it is assumed that collective farm sales fall proportionately. Prior to the adjustment, the total receipts of agriculture from collective farm sales amounted to 16.0 billion rubles. (These are the receipts from collective farm sales to households. In addition there were collective farm market sales of 1.8 billion rubles to institutions. To simplify the calculation it is assumed that, so far as concerns the values of the goods and services allocated to different use categories, the incidence of the increase in procurement prices needed to compensate the farmers for the decrease in the value of collective farm market sales to institutions would be offset by the corresponding reduction in the costs to the institutions of the goods obtained in the collective farm market.) Thus the decline in retail prices involves a reduction in farm income by $k \times 16.0$ billion rubles. Under our assumptions, there must be an increase in the procurement value of farm products by this same amount, over and above the amount due to elimination of subsidies.

(iv) The net reduction in the value of retail sales in state and cooperative shops after all adjustments must equal the difference between the amount of the turnover tax, in excess of subsidies, borne by these sales prior to the adjustment, i.e., 55.9 billion rubles, and the share of the additional compensatory increase in procurement prices that is absorbed by these sales as a result of the adjustment now under consideration, i.e., $.727 \times k \times 16.0$. Hence we have the equation:

$$k \times 111.5 \text{ billion rubles} = 55.9 \text{ billion rubles}$$
$$- .727 \times k \times 16.0 \text{ billion rubles}$$

or solving

$$k = 45.4 \text{ percent}$$

and for the necessary additional compensatory increase in procurement values, over and above that due directly to the elimination of subsidies, we have

$$.454 \times 16.0 \text{ billion rubles} = 7.3 \text{ billion rubles}$$

(v) This last sum is now allocated by use categories in the proportions indicated in (i), above.

(vi) In the case of army subsistence, an adjustment is made both under household consumption and under defense. But in accord with our assumptions, military pay is now revised downward by the same amount as army subsistence is increased.

(vii) In the initial calculation of the national product at established prices, farm income in kind is valued at an average of prevailing procurement and collective farm market prices, net of turnover taxes and gross of subsidies. Evidently, these are also the prices that prevail for farm produce after the elimination of taxes and subsidies. Furthermore, in terms of their effects on farm income the decrease in collective farm market prices and the additional compensatory increase in procurement prices are equivalent magnitudes, and thus apparently the average of realized farm prices is unchanged by this final adjustment. Accordingly, no change is now called for in farm income in kind.

# APPENDIX F. SOVIET OFFICIAL INCOME STATISTICS IN CURRENT RUBLES

It was mentioned at the outset that the Soviet government has published, in addition to its calculations in 1926–27 ruble prices, some percentage breakdowns of the national income in terms of current rubles. From various standpoints, it may be of interest to compare such of the latter as have been issued for 1937 with corresponding data based on our national economic accounts.

For the year 1937, the Soviet government has published the following data on the national income of the USSR in terms of current rubles: a tabulation of the disposition of the national income as between "consumption," "investment," and "reserves"; and a summary tabulation of the national income by economic sectors. These two tabulations are published only in percentage terms.

As with the concept of net national product used in this study, national income in the Soviet sense includes indirect taxes.[1] Also subsidies are apparently deducted; in other words, national income is valued at actual transfer prices, i.e., factor costs (including profits before subsidized losses) plus turnover taxes less subsidies.[2] As has been mentioned, however, it does not include a variety of services, and this omission must be taken into account in comparing the Soviet data with our own.

The Soviet official data on the disposition of national income in 1937 in terms of current rubles are as follows: consumption, 75.5 percent; investment, 21.6 percent; and reserves, 2.9 percent.[3]

If, on the basis of very rough estimates, the data in terms of prevailing rubles in Table 4 are adjusted to exclude "services" in the Soviet sense, the following more or less comparable tabulation is obtained.[4]

[1] D. I. Chernomordik, ed., *Narodnyi dokhod SSSR*, pp. 70, 84 ff.

[2] *Ibid.*, pp. 82–84.

[3] M. V. Kolganov, ed., *Narodnyi dokhod SSSR* (National Income of the USSR), p. 108.

[4] On the basis of data in Appendix Table 3, p. 136, the services to be omitted from the various use categories are taken to be the following:

|  | BILLION RUBLES |
|---|---|
| Housing; services | 19.9 |
| Trade union and other dues | 1.1 |
| Government administration, NKVD, personal services only | 6.0 |
| Communal services, personal services only | 13.7 |
| Defense, troop pay | 1.5 |
| Defense, army subsistence | 2.5 |
| Total | 44.7 |

|                                                                    | BILLION RUBLES | PERCENT [5] |
|--------------------------------------------------------------------|----------------|-------------|
| Consumption, excluding housing, services, trade union and other dues | 162.5          | 67.3        |
| Communal services, government, NKVD: outlays on commodities only   | 15.1           | 6.3         |
| Net investments                                                    | 50.3           | 20.9        |
| Defense: munitions only                                            | 13.5           | 5.6         |
| Net national product                                               | 241.3          | 100.0       |

In the absence of any explanation of the Soviet categories of consumption, investment, and reserves, the precise relation between the Soviet and our own tabulation must remain conjectural. If it is assumed that our "net investment" and "defense" taken together correspond to the Soviet "investment" and "reserves," the results nearly correspond, there being a divergence of only 2.0 percent of the net national product. Probably the discrepancy would be somewhat greater if it were not for a difference in the valuation of farm income in kind,[6] but all things considered, the agreement is striking.

The Soviet tabulation of national income by industrial sector and the corresponding one compiled here, again after the omission of services, are shown below.[7]

[5] The discrepancy between the indicated total and the sum of the indicated items is due to rounding.

[6] While farm income in kind is valued in the present study at average realized farm prices net of turnover taxes but gross of subsidies, it probably is valued in the Soviet calculations at these same prices net of both turnover taxes and subsidies. See Chernomordik, pp. 131 ff.

[7] The category "Services, including government" in Table 5 comes to much though not quite the same thing as the category of "services" that is omitted from the Soviet official national income concept. It will be noted, however, that the category "Services, including government" is indicated in Table 5 to have had an income of 38.0 billion rubles, whereas adjusting our calculation of national income by use to the Soviet concept we omitted from the net national product "services" having a value of 44.7 billion rubles. This discrepancy is to be explained in terms of the inclusion in the latter "services" of two items, passenger transportation and local utilities, which in Table 5 are classified under other headings than "Services, including government"; and presumably also by reference to the sizable statistical discrepancy in Table 5, which indicates the undervaluation of one or another of the economic sectors listed. It remains to observe that of the two items just referred to, passenger transportation is omitted from, and apparently local utilities (gas, electric, etc.) are included in, national income in the Soviet sense. Also it should be noted that in the case of finance, the Soviet procedure, in contrast with ours, seems to be to include interest paid as a part of the national income produced by the paying sector and to omit from national income the value of the services of finance as such.

| | | NET NATIONAL PRODUCT, SOVIET CONCEPT | |
| | | BERGSON DATA | |
| | SOVIET DATA,[8] PERCENT | Billions of rubles | Percent |
|---|---|---|---|
| Agriculture | 25.7 | 60.7 | 24.5 |
| Industry and construction | 58.9 | 121.2(133.2) | 48.9(53.7) |
| Trade, including restaurants | 9.5 | 37.5(25.5) | 15.1(10.3) |
| Finance | . . [9] | 1.9 | .8 |
| Transportation, communications, and other | 5.9 | 17.4 | 7.0 |
| Statistical discrepancy | . . | 9.4 | 3.8 |
| Total | 100.0 | 248.0 | 100.0 |

Again there seems to be a fair agreement in results, particularly in the case of the shares of agriculture, transport, and communications. There is at hand, furthermore, a plausible explanation for part of the observed discrepancies in the shares of industry and trade. In our calculation, a major element in the income produced by trade is the turnover tax paid by procurement agencies, amounting in all to 24.11 billion rubles. These agencies, however, are engaged not only in trading operations but also to some extent in the processing of the produce that they procure. For this reason, it is very likely that in the Soviet calculation some part of the procurement taxes has been tabulated under the heading of industry. The figures in parentheses in our tabulation indicate what the shares of industry and trade would be if the procurement taxes were divided equally between these two sectors. In view of the fact that this leads in the case of trade to nearly complete agreement between our own and the Soviet tabulation, one is led to assume that in the case of industry the remaining difference in results is to be explained in terms of the statistical discrepancy that appears in our tabulation.

Insofar as there is agreement, there is some reassurance as to the accuracy of both the Soviet and our own tabulations. The foregoing check may be of interest also insofar as it sheds light on the nature and scope of the Soviet income concepts and categories. These until now have not been entirely clear.

[8] M. V. Kolganov, p. 77.
[9] See above, note 7.

# APPENDIX G. COLLECTIVE FARM
# EMPLOYMENT AND EARNINGS IN 1937

As calculated here, collective farm employment in 1937, 32.0 million 280-day man years, represents the sum of:

|  | MILLIONS OF 280-DAY MAN YEARS |
|---|---|
| (i) Employment of able-bodied collective farmers, 16–59 years, on the collective farm and homestead | 26.7 |
| (ii) Employment of able-bodied collective farmers, 16–59 years, off the farm | 2.3 |
| (iii) Employment of collective farmers, other ages | 3.0 |

The first two figures on employment of collective farmers of 16–59 years are from B. Babynin. "Trudovye resursy kolkhozov i ikh ispol'zovanie" (Labor Resources of the Collective Farms and their Utilization), *Problemy ekonomiki*, 1940, No. 2, pp. 70, 74. The figure on employment in other age groups is calculated from Babynin's data together with other information in I. V. Sautin, *Proizvoditel'nost' i ispol'zovanie truda v kolkhozakh vo vtoroi piatiletke* (Productivity and Utilization of Labor in the Collective Farms during the Second Five Year Plan), p. 126.

Babynin's data on collective farm employment apparently are based on two kinds of Soviet collective farm labor statistics: global figures on the collective farm labor force and on the number of "labor days" credited to collective farmers for their collective farm work, obtained from collective farm annual reports; and data on the average number of labor days worked per calendar day (it will be recalled that the labor day is an accounting rather than calendar unit) and on the division of time between work on the collective farm and on the homestead, obtained from sample studies. For discussions of these basic data, see B. Babynin, "O balanse truda v kolkhozakh" (On the Labor Balance of the Collective Farms), *Planovoe khoziaistvo*, 1938, No. 12; M. Sonim, *Voprosy balansa rabochei sily* (Problems of the Balance of the Labor Force), Ch. IV. Elsewhere Babynin informs us that in the global figures based on annual reports the collective farm labor force may be "somewhat understated" ("Trudovye resursy . . . ," p. 70). Sonim (pp. 87–88) makes clear that this is actually the case; according to data he cites, the understatement for persons in the same age group is 5 to 15 percent. Presumably there would also be an understatement, though not necessarily in the same amount, in the data on employment calculated from these global figures on the labor force.

The employment data that have been cited are in terms of 280-day man years. This is a theoretic full-time man year. From data in Babynin, "Trudovye resursy . . . ," pp. 70 ff., it may be estimated that able-bodied male collective farmers of ages 16–59 years worked an average of some 270 days on and off the farm in 1937; and male and female collective farmers together, about 235 days.

The total income of collective farmers from all sources is computed as the following sum:

BILLION RUBLES

| | | |
|---|---|---|
| (i) | Money payments to collective farmers on a labor day basis; salaries of collective farm executives, premiums | 7.3 |
| (ii) | Net money income from sale of farm produce | 12.9 |
| (iii) | Net income in kind at farm prices | 29.4 |
| (iv) | Earnings from outside work | 6.7 |
| (v) | Total income | 56.3 |

The different elements in this total are obtained as follows:

(i) Money payments to collective farmers on a labor day basis, etc. See Appendix A, p. 103.

(ii) Net money income from sale of farm produce. According to Appendix A, p. 103, net money income from sales of farm products amounted to 14.2 billion rubles for all households. On the basis of the calculation made below, of the total net produce after production expenses in the hands of all farm households (including the produce of independent peasants and workers' gardens), the share in the hands of collective farm households was 90.5 percent. It is assumed that sales on the collective farm market are divided between the collective farm and other households in the same proportions.

(iii) Net income in kind at farm prices. According to Appendix A, p. 105, the net income in kind of all farm households amounted to 32.5 billion rubles. This sum likewise is assumed to be divided between collective farm and other households in proportion to their respective shares in the net farm produce at the disposal of farm households.

(iv) Earnings from outside work. On the basis of data in Nesmii, "Dokhody kolkhozov i kolkhoznikov," p. 100, the money income of collective farmers from services performed outside the collective farm are estimated at 361 rubles per household, or a total of 6.7 billion rubles for all 18.5 million households.

The calculation of the respective shares of the collective farm and other farm households in the total net produce at their disposal is as follows:

(i) According to TSUNKHU, "Sotsialisticheskoe sel'skoe khoziaistvo," p. 160, the shares of the gross output of agriculture produced by the different types of farm organizations in 1937 were respectively:

| | |
|---|---|
| State farms | 9.3 percent |
| Collective farms | 62.9 |
| Collective farm homesteads | 21.5 |
| Workers' gardens | 4.8 |
| Independent peasants | 1.5 |
| Total | 100.0 |

(ii) It is assumed that 25 percent of the gross output of collective farms is distributed in kind to collective farm households. This assumption, which is believed to be reasonably reliable, is based principally on the following information: (a) data on the share of the collective farm output of grain and legumes and of different animal products distributed to member households, in A. Baykov, p. 311, and in Nesmii, "Dokhody kolkhozov i kolkhoznikov," p. 86; and (b) data on the gross collective farm output of different crops, in TSUNKHU, "Sotsialisticheskoe sel'skoe khoziaistvo," p. 161.

(iii) On the basis of (i) and (ii) it is calculated that the collective farm payments in kind to the collective farm households amounted to 15.7 percent of the gross agricultural output (25 percent of 62.9 percent).

(iv) According to Appendix A, p. 105, the production expenses of agriculture constituted 35 percent of the gross harvest. For the purposes of the calculation, it is assumed that this ratio applies to the collective farm homesteads and workers' gardens.

(v) From (i) and (iv) it is computed that the *net* harvest of collective farm homesteads amounted to 14.0 percent (i.e., 65 percent of 21.5 percent) of the gross output of agriculture, and that the net harvest of workers' gardens and independent peasant homesteads amounted to 3.1 percent (i.e., 65 percent of 4.8 percent) of the gross output of agriculture.

(vi) Using the information in (iii) and (v), we obtain the following tabulations:

|  | PERCENT OF GROSS AGRICULTURAL OUTPUT OF ORGANIZATIONS AND HOUSEHOLDS | PERCENT OF NET PRODUCE AT DISPOSAL OF HOUSEHOLDS |
|---|---|---|
| Collective farm payments in kind to collective farm households | 15.7 | |
| Net harvest of collective farm homesteads | 14.0 | |
| Net produce at disposal of collective farm households | 29.7 | 90.5 |
| Net harvest of workers' gardens and independent peasant homesteads | 3.1 | 9.5 |
| Net produce at disposal of households | 32.8 | 100.0 |

# PUBLICATIONS CITED

## ABBREVIATIONS

Gosplan      *Gosudarstvennaia Planovaia Komissiia* (State Planning Com-
             mission)
Narkomfin    *Narodnyi Komissariat Finansov SSSR* (People's Commissariat
             of Finances of the USSR)
TSUNKHU      *Tsentral'noe Upravlenie Narodno-Khoziaistvennogo Uchëta*
             (Central Administration of National Economic Accounting)

# PUBLICATIONS CITED, TOGETHER WITH
# TRANSLATIONS OF RUSSIAN TITLES

Arnold, A. Z. Banks, Credit and Money in Soviet Russia. New York, 1937.
Babynin, B. "O balanse truda v kolkhozakh" (On the Labor Balance of the Collective Farms), *Planovoe khoziaistvo*, 1938, No. 12.
—— "Trudovye resursy kolkhozov i ikh ispol'zovanie" (Labor Resources of the Collective Farms and Their Utilization), *Problemy ekonomiki*, 1940, No. 2.
Baran, Paul A. "National Income and Product of the USSR in 1940," *Review of Economic Statistics*, November, 1947.
Baster, Nancy. Agrarian Overpopulation in the USSR, 1921–1940. Unpublished essay on file at Russian Institute, Columbia University, 1949. Pp. 84.
Baykov, A. The Development of the Soviet Economic System. Cambridge, England, 1946.
Bergson, Abram. "Comments," in "The Economy of the USSR"; "Papers and Proceedings of the Fifty-ninth Annual Meeting of the American Economic Association," *American Economic Review*, 1947, No. 2 (May).
—— "A Problem in Soviet Statistics," *Review of Economic Statistics*, November, 1947.
—— "Socialist Economics" in H. Ellis, A Survey of Contemporary Economics. Philadelphia, 1948.
—— "Soviet National Income and Product in 1937," *seriatim*, Part I, "National Economic Accounts in Current Rubles," *Quarterly Journal of Economics*, Vol. 64, No. 2 (May, 1950); and Part II, "Ruble Prices and the Valuation Problem," *ibid.*, No. 3 (August, 1950).
—— The Structure of Soviet Wages. Cambridge, 1944.
Bienstock, G., S. Schwartz, and A. Yugow. Management in Russian Industry and Agriculture. New York, 1944.
Bogolepov, M. "Finansy SSSR nakanune tret'ei piatiletki" (Finances of the USSR on the Eve of the Third Five Year Plan), *Planovoe khoziaistvo*, 1937, No. 3.
Central Administration of Economic and Social Statistics. (See TSUNKHU.)
Cherniavskii, V., and S. Krivetskii. "Pokupatel'nye fondy naseleniia i roznichnyi tovarooborot" (Purchasing Power of the Population and the Retail Turnover), *Planovoe khoziaistvo*, 1936, No. 6.
Chernomordik, D. I., ed., Narodnyi dokhod SSSR (National Income of the USSR). Moscow, 1939.
Clark, Colin. A Critique of Russian Statistics. London, 1939.
—— "Russian Income and Production Statistics," *Review of Economic Statistics*, November, 1947.
Colm, Gerhard. "Experiences in the Use of Social Accounting in Public Policy in the United States," International Association for Income and Wealth, Income and Wealth, Series I, Cambridge, England, 1951.

Dallin, D. J. and B. I. Nicolaevsky. Forced Labor in Soviet Russia. New Haven, 1947.

Dobb, Maurice. "Further Appraisals of Russian Economic Statistics," *Review of Economic Statistics,* February, 1948.

—— Soviet Economic Development since 1917. London, 1948.

—— Soviet Economy and the War. New York, 1943.

D'iachenko, V. P., ed. Finansy i kredit SSSR (Finances and Credit of the USSR). Moscow, 1938.

*Finansovoe i khoziaistvennoe zakonodatel'stvo* (Financial and Economic Legislation).

*Finansovyi i khoziaistvennyi biulleten'* (Financial and Economic Bulletin).

Gerschenkron, A. "Comments on Naum Jasny's 'Soviet Statistics,'" *Review of Economics and Statistics,* August, 1950.

—— A Dollar Index of Soviet Machinery Output, 1927–28 to 1937. Santa Monica, California: The RAND Corporation, 1951.

—— Economic Relations with the USSR. New York, 1945.

—— "The Soviet Indices of Industrial Production," *Review of Economic Statistics,* November, 1947.

Glezin, S. (compiler). Biudzhetnaia sistema Soiuza SSR (Budget System of the USSR). Moscow, 1937.

Gordon, Manya. Workers Before and After Lenin. New York, 1941.

Gosplan. Narodno-khoziaistvennyi plan na 1936 god (National Economic Plan for 1936). Second edition. Moscow, 1936.

—— Summary of Fulfillment of the First Five Year Plan. Moscow, 1933.

—— Tretii piatiletnii plan razvitiia narodnogo khoziaistva Soiuza SSR, 1938–42 (Third Five Year Plan of Development of the National Economy of the USSR, 1938–42). Moscow, 1939.

Gosudarstvennyi plan razvitiia narodnogo khoziaistva SSSR na 1941 god (State Plan for the Growth of the National Economy of the USSR for 1941). This volume was issued officially as an appendix to a decree of the Council of People's Commissars and the Central Committee of the Communist Party, January 17, 1941. It has been reissued by the American Council of Learned Societies.

Granick, David. Plant Management in the Soviet Industrial System. Doctoral dissertation, Columbia University, 1951.

Grin'ko, G. F. Finansovaia programma Soiuza SSR na 1937 god (Financial Program of the USSR for 1937). Moscow, 1937.

Hicks, J. R. "The Valuation of Social Income," *Economica,* May, 1940.

—— "The Valuation of Social Income," *Economica,* August, 1948.

Hubbard, L. E. The Economics of Soviet Agriculture. London, 1939.

—— Soviet Trade and Distribution. London, 1938.

Hutt, W. H. "Two Studies in the Statistics of Russia," *The South African Journal of Economics,* March, 1945.

Jasny, Naum. "Intricacies of Russian National Income Indexes," *Journal of Political Economy,* August, 1947.

—— "Results of the Five Year Plans," in W. Gurian, ed., The Soviet Union. Notre Dame, Indiana, 1951.

—— The Socialized Agriculture of the USSR. Stanford, 1949.

—— The Soviet Economy during the Plan Era. Stanford, 1951.

—— The Soviet Price System. Stanford, 1951.

—— Soviet Prices of Producers' Goods. Stanford, 1952.

—— "Soviet Statistics," *Review of Economics and Statistics,* February, 1950.

Kaldor, N. "The 1941 White Paper on National Income and Expenditure," *Economic Journal,* June–September, 1942.

Khromov, P. A. Amortizatsiia v promyshlennosti SSSR (Depreciation in Soviet Industry). Moscow, 1939.

Kolganov, M. V., ed. Narodnyi dokhod SSSR (National Income of the USSR). Moscow, 1940.

Kournakoff, S. N. Russia's Fighting Forces. New York, 1942.

Kozlov, G. A., ed. Finansy i kredit SSSR (Finances and Credit of the USSR). Moscow, 1940.

—— Khoziaistvennyi raschët v sotsialisticheskom obshchestve (Economic Accounting in Socialist Society). Moscow, 1945.

Kutler, K. N. Gosudarstvennye dokhody SSSR (Government Incomes of the USSR). Leningrad, 1940.

Kutyrev, S. M. Analis balansa dokhodov i raskhodov khoziaistvennoi organizatsii (Analysis of the Balance of Incomes and Outlays of the Economic Organization). Moscow, 1948.

Kuznets, Simon. National Product since 1869. New York, 1946.

—— "On the Valuation of Social Income," *Economica,* May, 1948.

Kuznetsov, G. I. Otpusknye i roznichnye tseny i torgovye nakidki na promtovary (Wholesale and Retail Prices and Trading Margins on Industrial Goods). Moscow, 1936.

—— ed. Sbornik otpusknykh i roznichnykh tsen i torgovykh nakidok na prodovl'stvennye tovary (Handbook of Wholesale and Retail Prices and Trading Margins on Food Products). Moscow, 1936.

The Land of Socialism Today and Tomorrow. (Reports and Speeches at the Eighteenth Congress of the Communist Party . . . March 10–21, 1939.) Moscow, 1939.

Laptev, I. "Kolkhoznye dokhody i diferentsial'naia renta" (Collective Farm Income and Differential Rent), *Bolshevik,* August, 1944. A translation appears in the *American Review on the Soviet Union,* May, 1945.

Lenin, V. I. Selected Writings. New York, 1942.

Levin, D. C., and M. G. Poliakov. Kal'kulirovanie sebestoimosti produktsii miasnoi promyshlennosti (Calculation of Costs of Production of the Meat Industry). Moscow, 1936.

Lorimer, Frank. The Population of the Soviet Union. Geneva, 1946.

Lozovsky, A. Handbook on the Soviet Trade Unions. Moscow, 1937.

Margolin, N. S. Balans denezhnykh dokhodov i raskhodov naseleniia (Balance of the Monetary Incomes and Expenditures of the Population). Moscow-Leningrad, 1940.

—— Voprosy balansa denezhnykh dokhodov i raskhodov naseleniia (Problems of the Balance of Money Incomes and Expenditures of the Population.) Moscow-Leningrad, 1939.

Narkomfin. Alfavitnyi perechen' promtovarov po stavkam naloga s oborota i biudzhetnykh natsenok (Alphabetic List of Industrial Commodities by Rates of the Turnover Tax and Budget Surcharges). Moscow, 1938.

—— Gosudarstvennyi biudzhet SSSR za vtoruiu piatiletku, 1933–37 gg. (State Budget of the USSR for the Second Five Year Plan, 1933–37). Moscow, 1939.

—— Klassifikatsiia raskhodov i dokhodov edinogo gosudarstvennogo biudzheta na 1936 g. (Classification of the Expenditures and Revenues of the Unified State Budget for 1936). Moscow, 1935.

—— Otchët ob ispolnenii gosudarstvennogo biudzheta SSSR za 1935 (Account on the Fulfillment of the Government Budget of the USSR for 1935). Moscow, 1937.

Nesmii, M. "Dokhody kolkhozov i kolkhoznikov" (Incomes of Collective Farms and Farmers), *Planovoe khoziaistvo,* 1938, No. 9.

—— "Finansovoe khoziaistvo kolkhozov" (Financial Economy of the Collective Farms), *Planovoe khoziaistvo,* 1939, No. 8.

*New Republic,* Supplement, May 16, 1949.

*New York Times.* October 17, 1948.

Nikolaev, M. V. Bukhgalterskii uchët (Accounting). Moscow, 1936.

Petrov, V., and V. Fisherov (compilers). Nalog s oborota (The Turnover Tax). Moscow, 1936.

*Planovoe khoziaistvo.*

Plotnikov, K. N. Biudzhet Sovetskogo gosudarstva (Budget of the Soviet Government). Moscow, 1945.

Prokopovicz, S. N. *Quarterly Bulletin of Soviet Russian Economics.*

Rovinskii, N. N. Gosudarstvennyi biudzhet SSSR (Government Budget of the USSR). Moscow, 1939, 1944, 1949.

Samuelson, P. A. "Evaluation of Real National Income," Oxford Economic Papers, Vol. II, No. 1 (January, 1950).

Sautin, I. V. Kolkhozy vo vtoroi Stalinskoi piatiletke (Collective Farms in the Second Stalinist Five Year Plan). Moscow, 1939.

—— Proizvoditel'nost' i ispol'zovanie truda v kolkhozakh vo vtoroi piatiletke (Productivity and Utilization of Labor on the Collective Farms during the Second Five Year Plan). Moscow, 1939.

Schwartz, Harry. "A Critique of 'Appraisals of Russian Economic Statistics,'" *Review of Economic Statistics,* February, 1948.

Second Session of the Supreme Soviet of the USSR. Moscow: August, 1938.

Smilga, A. "Finansy sotsialisticheskogo gosudarstva" (Finances of the Socialist Government), *Problemy ekonomiki,* 1937, No. 2.

Sobranie Zakonov i Rasporiazhenii SSSR (Collected Laws and Decrees of the USSR).

Sokolov, N. "Bor'ba Gosbanka za khozraschët v tret'ei piatiletke" (Struggle of the State Bank for Economic Accounting in the Third Five Year Plan), *Planovoe khoziaistvo,* 1939, No. 6.

Sonim, M. Voprosy balansa rabochei sily (Problems of the Balance of the Labor Force). Moscow, 1949.

Sonim, M., and B. Miroshchnichenko. Podbor i obuchenie rabochikh kadrov v promyshlennosti (Selection and Training of Labor Cadres in Industry). Moscow, 1944.

Stalin, J. Leninism: Selected Writings. New York, 1942.

State Planning Commission of the USSR. (See Gosplan.)

Studenski, Paul. "Methods of Estimating National Income in Soviet Russia," National Bureau of Economic Research, Conference on Research in Income and Wealth, Studies in Income and Wealth, Vol. VIII. New York, 1946.

Studenski, Paul, and Julius Wyler. "National Income Estimates of Soviet Russia," in "The Economy of the USSR"; "Papers and Proceedings of the Fifty-ninth Annual Meeting of the American Economic Association," *American Economic Review,* 1947, No. 2 (May).

Suchkov, A. K. Dokhody gosudarstvennogo biudzheta SSSR (Incomes of the Government Budget of the USSR). Moscow, 1945.

Tretiia Sessiia Verkhovnogo Soveta SSSR, 25–31 maia, 1939; stenograficheskii otchët (Third Session of the Supreme Soviet of the USSR, May 25–31, 1939; verbatim report). Moscow, 1939.

Trubnikov, S. "Istochniki komplektovaniia rabochei sily v SSSR" (Sources of Recruitment of the Labor Force of the USSR), *Problemy ekonomiki,* 1936, No. 6.

Tsentral'noe Statisticheskoe Upravlenie (Central Statistical Administration). Kontrol'nye tsifry, 1928–29 (Control figures, 1928–29). Moscow, 1929.

——— Statisticheskii spravochnik SSSR za 1928 (Statistical Handbook of the USSR for 1928). Moscow, 1929.

TSUNKHU. Chislennost' i zarabotnaia plata rabochikh i sluzhashchikh v SSSR (Number and Wages of Wage Earners and Salaried Workers in the USSR). Moscow, 1936.

——— Kolkhoznaia i individual'no-krestianskaia torgovlia (Collective Farm and Independent Peasant Trade). Moscow, 1936.

——— Socialist Construction in the USSR. Moscow, 1936.

——— "Sotsialisticheskoe sel'skoe khoziaistvo Soiuza SSR" (Socialist Agriculture of the USSR), *Planovoe khoziaistvo,* 1939, No. 7.

——— "Sotsialisticheskoe stroitel'stvo Soiuza SSR" (Socialist Construction of the USSR), *Planovoe khoziaistvo,* 1939, No. 8.

——— Trud v SSSR (Labor in the USSR). Moscow, 1936.

Turgeon, L. "On the Reliability of Soviet Statistics," *Review of Economics and Statistics,* February, 1952.

United States Department of Commerce. "National Income," Supplement to *Survey of Current Business,* July, 1947.

United States National Resources Committee. The Structure of the American Economy, Part I. Washington, D.C., June, 1939.

Vilenskii, L. "Finansovye voprosy promyshlennosti" (Financial Problems of Industry), *Planovoe khoziaistvo,* 1938, No. 10.

Voznesenskii, N. (Voznesensky, N.) Voennaia ekonomika SSSR v period otechestvennoi voiny (War Economy of the USSR in the Period of the Patriotic War). Moscow, 1947.

—— The Growing Prosperity of the Soviet Union. New York, 1941.

Wyler, Julius. "The National Income of Soviet Russia," *Social Research,* December, 1946.